THE MIRROR AND THE MIND

PRINCETON MODERN KNOWLEDGE

Michael D. Gordin, Princeton University, Series Editor

For a list of titles in the series, go to https://press.princeton.edu/series
/princeton-modern-knowledge.

The Mirror and the Mind

A HISTORY OF SELF-RECOGNITION
IN THE HUMAN SCIENCES

KATJA GUENTHER

PRINCETON UNIVERSITY PRESS

PRINCETON & OXFORD

Published by Princeton University Press
41 William Street, Princeton, New Jersey 08540
99 Banbury Road, Oxford OX2 6JX

press.princeton.edu

All Rights Reserved

ISBN 978-0-691-23725-1
ISBN (e-book) 978-0-691-23726-8

Library of Congress Control Number: 2022940842

British Library Cataloging-in-Publication Data is available

Editorial: Eric Crahan and Barbara Shi
Production Editorial: Karen Carter
Jacket/Cover Design: Karl Spurzem
Production: Danielle Amatucci
Publicity: Kate Farquhar-Thomson and Alyssa Sanford
Copyeditor: Kathleen Kageff

This book has been composed in Arno

Printed on acid-free paper. ∞

Printed in the United States of America

10 9 8 7 6 5 4 3 2 1

To my family

Strange, that there are dreams, that there are mirrors.
Strange that the ordinary, worn-out ways
of every day encompass the imagined
and endless universe woven by reflections.

—JORGE LUIS BORGES, "MIRRORS"

CONTENTS

THE MIRROR AND THE MIND

Introduction

THE MIRROR SELF-RECOGNITION TEST is, in its experimental setup at least, extremely simple. A mirror is placed in an open space, often attached to a wall or standing vertically on the floor. The subject, perhaps a small child or an animal, is put in front of it. These circumstances are reproduced daily in countless homes around the world. Nothing could be more common. And yet, if present at the right time, paying attention to the right subject, one might observe something extraordinary. The subject might start by acting surprised at the image. It might approach it cautiously, or perhaps even aggressively, because the image seems to be similarly suspicious and unfriendly. But often, and after a little time, things begin to change. The subject no longer appears on edge. It is relaxed, happy even; perhaps it smiles. Playfulness has replaced suspicion. The subject might move its hands, its eyes shifting back and forth between the physical body and its reflection. It might open its mouth wide, leaning into the mirror so that it can see its teeth in the reflection, perhaps even pick out a piece of food caught between two teeth. Throughout this process, we are confronted only with outward behavior, the way the body moves at different stages of the encounter. But it is easy, unavoidable perhaps, to see these as the outward signs of an internal psychological drama, whereby the subject looks in the mirror and slowly comes to see itself. Might we be witness to the dawning of self-consciousness?

When I told people that I was working on a book about mirrors, I received a wide range of responses. A medievalist colleague cited Saint Paul: "Videmus nunc per speculum et in aenigmate" (1 Cor. 13:12, Vulg.) (We do not now see [God] but through a mirror, darkly). Another confessed to me that he used to stare at the mirror as a young man to see if he really existed (he did not say what was the result). One shared how intrigued she was with the use of mirrors in the film *The Black Swan*, where the mirror revealed to the protagonist Nina Sayers (played by Natalie Portman) her hidden and dark identities. Yet others evoked the myth of Narcissus losing himself in his reflected image, or Aesop's dog, who foolishly jumped at his reflected image in a river to steal another dog's bone, thus losing the one he had to start with. My own motivation for the book came from a personal experience of observing my twin girls playing with a mirror when they were small. Despite its mundane ubiquity, the mirror remains a strange and endlessly fascinating object, because it seems to tell us new truths about who we are.

A History of Mirrors

The mirror has not always been an everyday item. In the ancient world, when mirrors were made from polished bronze or metal alloys, they were available only to a select few.[1] Their preciousness depended on the kind of metal chosen for their production. Egyptians and Sumerians made copper, bronze, gold, and silver mirrors. The Romans used polished obsidian, a black volcanic rock. What ancient mirrors had in common is that they were all fairly small, around five to eight inches in diameter. They shared this property with the earliest glass mirrors in the Middle Ages, which were constructed by applying a layer of metal to glass. Given the heat needed to melt the metal, and the challenge of blowing glass that was sufficiently flat, medieval glass mirrors were small and distorting, as in Jan van Eyck's famous *Arnolfini Portrait* (fig. o.1 and plate 1).[2]

A dramatic increase in size and decrease in cost was secured in the sixteenth century by glassmakers from the island of Murano near Venice. Drawing on centuries of glassmaking expertise (and using the highest-quality ingredients including seawater, and a type of wood that burned to produce a clear flame), they were able to make glass that was pure and clear. Thanks to their refined technique, they were also able to make larger mirrors, measuring up to forty square inches. These brought the Republic of Venice substantial wealth, and its wares were prized across Europe and the Middle East.[3] Accordingly, the Venetians guarded their secrets jealously—Murano workers were prohibited from emigrating, or even speaking with strangers.[4]

The French broke the Venetian monopoly in the late sixteenth century. The company Saint-Gobain, heavily subsidized by the state, managed to lure a few artisans from Murano to Paris, where they perfected the method of casting large glass mirrors.[5] For the next 150 years, French-cast glass mirrors set the

FIGURE 0.1. Jan van Eyck, *The Arnolfini Portrait*, 1434. The mirror also has a revealing function: two people are entering the room, one of whom could be the painter. Source: Wikimedia Commons.

standard, as represented most famously by the mirror hall at Versailles, which was unveiled to the public in 1682.[6] The French glassmakers used mercury to add a reflective coat to the glass, which took a serious toll on their health. In addition, the mercury left a greenish-gray tint that muddied the reflection. Mirror makers were thus on the lookout for alternatives, the most famous of which came from a procedure developed by the German chemist Justus von Liebig in 1856.[7] He used an aldehyde reaction (he found that aldehydes reduced silver salts to metallic silver) to add a layering of silver to glass.[8] Living through the "century of optical instruments and visualization,"[9] Liebig originally developed his method in order to improve mirrors used in scientific instruments, such as microscopes, telescopes, or Hermann von Helmholtz's famous ophthalmoscope (*Augenspiegel*). Although at first Liebig's method could not compete with the existing methods (the factory near the Bavarian city of Fürth to which Liebig sold his license had to close its doors after only two years of production), over the course of the nineteenth century, safety regulations came to restrict the use of mercury, and the Liebig method became dominant.[10] It also sped up the production process.[11] By the end of the nineteenth century, mirrors were everywhere: in shops (the development of special insurance against mirror breakage encouraged department stores to make heavy use of mirrors for interior decoration),[12] in cafés and foyers, and in almost every private home. Today the mirror has become common enough to be almost totally unremarkable.[13] It pervades the most intimate as much as the most public spaces. It has also, in the guise of the mirror test, pervaded the history of the modern mind sciences.

Psychology and Its Others

The history of the mind sciences has traditionally been told as a sequence of different intellectual movements.[14] The story often starts with Wilhelm Wundt's experimental introspection in Leipzig in the 1870s, which in the following decades made inroads into the United States. Wundt's students were deeply involved in building the infrastructure of American psychology. Stanley Hall founded the *American Journal of Psychology* in 1887 and, with Edward Scripture, the American Psychological Association (APA) in 1892. By the 1910s, however, some began to worry that Wundt's introspection leaned too heavily on subjective and thus unreliable experience. Most importantly, in his "behaviorist manifesto" from 1913, John B. Watson sought to bypass introspection entirely, focusing solely on external behavior. By studying how individual stimuli were tied to particular responses, psychology could work toward a goal of "prediction and control," which he thought would have wide-ranging social applications.[15] In the 1940s, so the story goes, behaviorism too came under assault. Building on the efforts of cyberneticians to promote a broad interdisciplinary conversation, as exemplified by the legendary Macy Conferences in New York (1946–53), developmental psychologists came together with specialists on the new electronic computers, with neuropsychologists, and with linguists in the Hixon Symposium at the California Institute of Technology in 1948 to seek ways to break free from the narrow confines of behaviorist science. By the 1960s their so-called cognitive revolution, which returned attention to the study of consciousness, had transformed the field. The mind could be studied scientifically after all, in its core functions such as memory, or in the study of language. This narrative has been complicated by a range of excellent studies, but its stagist structure and the narrative in its broad outlines persists.[16]

This book, by examining a single test, and following it wherever it appears, carves a less familiar path. The mirror test is particularly valuable because it allows us to think through the disciplinary richness of the mind sciences. For reasons that will become clear, the test often sat on the margins of different academic fields, connecting psychology with neurology, but also with evolutionary biology, psychoanalysis, anthropology, linguistics, and cybernetics. It also moved between pure and applied research, and research and therapy. It did not respect national borders either. As we will see, an analysis of the mirror test requires us to move between Germany, France, Britain, North America, and beyond.

Though it can be followed as a guiding thread crossing national and disciplinary lines, the mirror test did not form an intellectual tradition, in the sense of a clearly articulated network of textual references. Some mirror researchers appealed to a slowly expanding canon—Darwin and Preyer, and then Lacan, Amsterdam, and Gallup—and in the first two chapters of the book I will analyze the type of intellectual inheritance and influence with which most historians will be familiar. The later chapters, however, tend to deal with scientists who were mostly unaware of these predecessors, and worked independently. Often the mirror emerged in their work serendipitously. When mirrors are common household objects, one might easily encounter a mirror response by chance or integrate it into a research project as an ad hoc measure.

What motivated the turn to the mirror in these different instances was rather a common problematic. While most canonical psychology had focused on the male adult, mirror recognition seemed suited to the psychological study of infants and animals, with occasional forays into the realm of robotics. For this question, the dominant figure in the late nineteenth century was not Wundt but his compatriot Wilhelm Preyer. That is why, at times, we travel along what seem to be the backroads of the mind sciences, discovering figures and movements that have been ignored by much of the scholarship. In particular, this backroad travel draws our attention to numerous women—Milicent Shinn, Charlotte Bühler, Beulah Amsterdam, and Hilde Bruch—who fought against marginalization in their time and are often passed over today. When the mirror test did take paths that parallel and even flow into the mainstream, it encourages us to reconsider traditional narratives in ways that build on the previous scholarship. This strand of psychological research raised questions about the interpretation of behavior and the dangers of introspection, long before any putative beginnings of behaviorism. And yet, focused as it was on the emergence of higher functions, this research retained an interest in communication and the formation of concepts throughout in ways that would allow, in the 1960s, engagement with the "cognitive revolution."

The study of the nonlinguistic mind came to the fore in the final third of the nineteenth century, because it seemed a promising way to address a problem that had recently emerged but which had far-reaching consequences. For the past hundred years, the category of the human had been invested with enormous political significance. Political power was no longer supposed to derive from one's place in a larger social whole, a whole that was just one part of a divinely ordained "chain of being." Rather it was as "humans" (or, more often, "men") that citizens came together and demanded a role in their own governance. First, qua "men," people did not enter the social realm as members of a particular estate, or guild, or even class, but as individuals. Second, these individuals were granted authority through their reason. Considered the quintessential human faculty, reason was often used to justify the value of liberty and belief in progress and was declared as foundational in new declarations of rights and constitutions across the Atlantic world. For many, reason and individuality found their ultimate foundation in human language, which was considered to be qualitatively different from the forms of communication found among other animals.

Darwin's theory of evolution shook the foundations of this politics. This is not because he deconstructed the boundaries of the human. However much cartoonists lampooned Darwin the man-ape, there was no danger that the evolutionary theorist would come to confuse humans with other creatures. The shock of Darwin's work was rather that in placing humanity back into the bosom of nature, he seemed to undermine our exclusive hold on those properties that were essential to the new politics. How did the rules of natural selection fit with our dignity as individuals? Given the "descent of man," could it be plausibly suggested that human reason was qualitatively different from the mental powers of other animals? Crucially, Darwin denied that human language was sui generis.

In this context the mirror test became particularly attractive. It bypassed the question of language, because it could be used on nonhuman animals and children before they could speak, and it allowed scholars to reassert human superiority without falling back on religious notions or metaphysical claims about the soul. Not only did most think, for almost a hundred years, that humans were the only creatures able to recognize themselves in the mirror; they also held that mirror recognition demonstrated precisely the characteristics that were meant to set humans apart. First, in the mirror you identify yourself as an individual. It was assumed that other animals failed the test because they merely saw another animal. Second, it seemed to be the result of higher thinking. After all the mirror test required recognition, the application of a concept to an image. The mirror test, that is, did not simply serve as a shibboleth separating out humans from other creatures. It did so in a way that flattered humans and endorsed the image they had created for themselves. It offered a means to suture the gaping hole in human pride that had been opened up by the Darwinian revolution.

In the century spanning 1870 to 1970, mirror self-recognition became central to the definition of human specificity. Though other demarcators, most notably language, returned periodically, mirror self-recognition seemed to be the most reliable.[17] Not least it offered that peculiar advantage that it could be tested.[18] The mirror seemed to be able to produce experimental evidence of human distinctiveness, and scientists used it to show that humans and only humans were able to recognize their reflection.[19] Nevertheless, as a test, mirror self-recognition gained its authority by holding open the possibility of alternative answers, and researchers had to take seriously the prospect that animals might recognize themselves. There was always a chance that the test could dramatically change its meaning. Indeed, since 1970, as a Noah's ark of nonhuman animals have been shown to pass the test, it has become a favored tool of researchers arguing for animal rights. That perhaps points to one of the reasons why scientists have returned insistently to the mirror self-recognition test over the past 150 years: never simply a means to confirm existing theories, as an experimental system, it never lost the ability to surprise.[20]

The Mirror and Material Culture

Whatever hopes individuals might have invested in the mirror test in theory, in the heat of experimentation they were confronted with a set of difficulties resulting from the exigencies of material culture.[21] Historians of science have been interested in experimental systems and the materiality of scientific instruments for some time. Much in this work has focused on the physical sciences. Take, for instance, Peter Galison's *Image and Logic*, a book about the "machines of physics," such as the bubble chamber and the Geiger counter.[22] These machines allowed physicists to study the "microworld": the smallest forms of matter such as electrons, photons, protons, and quarks because the physical properties of those particles interacted with the machines to produce representations of their activity: for instance bubble chambers were constructed in such a way that subatomic particles would produce paths of small bubbles in superheated liquid hydrogen. Galison's machines could mediate between scientific objects and the knowledge about these objects because they could interact materially with those objects.[23]

Hans-Jörg Rheinberger's approach differs from Galison's. He posits a greater instability both at the level of machines (elements of his "experimental systems") and of the scientific objects (his "epistemic things"). By virtue of their capacity for "differential reproduction," experimental systems were capable of creating unforeseeable scientific events; they were "machine[s] for making the future."[24] But, as in Galison's account, the machines could help produce epistemic things because they shared physical properties. Rheinberger's main example is the in vitro synthesis of proteins. Of course these were "things embodying concepts";[25] the transfer RNA that emerged from "soluble RNA" within Rheinberger's experimental system embodied Francis Crick's "adaptor hypothesis," which introduced the language of information transfer. But as Rheinberger himself makes clear, his "epistemic things" were "material entities or processes—physical structures, chemical reactions, biological functions."[26] In these two canonical accounts of material culture, then, we see a commonality between the apparatuses used by scientists and the objects studied; *material* culture is useful for studying *material* things.[27]

In more recent years, the study of material culture has carved a path into the mind sciences by examining media, from Alessandro Mosso's brain plethysmography, through the EEG, to fMRI scans, but also photographs, film, and writing systems.[28] As Cornelius Borck has pointed out, in contrast to physiological inscriptions produced by the activity of other organs, like the electrocardiogram, the scribbles produced by activity of the brain in the EEG were considered to be not simply a *trace* produced by the human body, but forms of *writing*, produced by and thus revealing the subject.[29] Historians of science have shown how these "psychographies" were, in the tradition of Jules de Marey's *méthode graphique*, attractive as forms of nature writing itself—providing immediate and transparent access into the workings of the mind.[30]

The mirror too is a medium (though as we will see, what type of medium was a vexing question). Yet the mirror function is different from the function of other media. In most cases, the medium is the means by which the scientist comes to know the subject. It stands between the researcher and what is researched, controlling the flow of information between the two. But the mirror doesn't itself offer researchers anything new or different. In the self-recognition test, the scientist rarely, if ever, looks at the reflection. Rather it is the *subject* who is gazing into the mirror, and the scientist is concerned with how that subject responds.

At the most basic level, the mirror is an apparatus that alters the path of light rays in a consistent way, so as to produce the illusion that the reflected object exists in another position in space. Though the image is inverted, this is usually apparent only when the mirror reflects the written word, and in most cases what we see is not dissimilar from what we are able to perceive without a mirror.[31] That is why it is so easy to mistake a mirror image for the real thing. A mirror is perhaps useful for observing an object from an otherwise inaccessible angle, but normally there is nothing surprising or novel about the image it offers.

Something different occurs, however, when we turn our attention to the reflection of our own bodies. For then, we see more than another body in space. We see an object that is both familiar and unfamiliar, what is closest to us viewed from a perspective that we are normally denied. In the mirror we see ourselves as if through the eyes of another. In transforming our bodies from something that we are and feel into something that we encounter as a distinct and separate object, the mirror prompts a range of otherwise unusual cognitive acts. It encourages us to project our proprioceptive selves into the external world. Consequently, the mirror image could become a vehicle for a range of subjective impressions—in the mirror we see our desires and our fears—and even higher-order concepts, such as the self or ego. And this was especially important for those subjects, such as infants and animals, who had often been excluded from studies of higher function, because they lacked the linguistic capacity required for the tests. Here, rather than being a more or less transparent point of access through which the scientist is able to study the mind, the mind is revealed through the very distortions produced when the subject sees its own mirror reflection.

And yet, these distortions are not immediately available to the researchers. Because the mirror test most often involved non- or prelinguistic subjects, one could not ask them to describe their experience. Instead, scientists were reliant on an examination and interpretation of the subjects' behaviors. The focus on behavior helps drive one of the central questions of the mirror self-recognition test. How do we actually know what the subject sees in the mirror? On the one hand, interpretation seems easy. When we as adult humans look in the mirror, we see a reflection of our own bodies, and so it is only natural to look for the same recognition in other creatures. An aggressive response might suggest that the subject takes the image to be someone else. A smile, perhaps, is an indication that something has clicked, that the subject has recognized itself. But on the other hand, however intuitive these interpretations were, they proved extremely difficult to justify.

That is why the mirror self-recognition test cannot be considered in its simplicity. As we will see, scientists tended to build around it a range of practices and techniques to control and order the ambiguity of the behavior. They developed notation strategies; compiled strict testing protocols, often laid out in questionnaires; and, when it became available, were early adopters of video recording. The study of the mirror test thus fits into another strand of literature on material culture concerning inscription practices.[32] Ever since Ursula Klein described Berzelian formulas as "paper tools" by which chemists could grapple with and thus come to know chemical reactions, historians have extended attention away from the site of experiment to understand the production of knowledge.[33] More recently, scholars such as Anke te Heesen, Andrew Mendelsohn, and Volker Hess have shown that "paper technologies"—notebooks and medical case histories—were used to organize different kinds of information, in ways that opened up science to the everyday.[34] In this way, scholars have expanded the reach of what counts as scientific material culture: from scientific models to "things that talk," which includes objects as diverse as glass flowers, Rorschach tests, and soap bubbles.[35]

Mirror researchers sought to control the ambiguity of the mirror encounter in another way: by folding it into a range of neighboring but distinct scientific theories, most importantly neurology and its neuroscientific heirs, but also anthropology and linguistics. These disciplines provided an authoritative scientific basis for interpretations of mirror behavior. For example, based on the knowledge that child brain development involved the building of associations between various sensory and motor centers, mirror researchers were inclined to see the effects of those associations in mirror behavior, perhaps even to mark the moment when particular connections were made.

Plan of the Book

This book will approach the history of mirror experiments by foregrounding two, broadly speaking consecutive, problematics. In part 1 ("Identifications"), I examine the history of the mirror test from the late eighteenth century until around 1970, though focusing mostly on the final hundred years. For researchers at this time, the central problem of the mirror test was determining whether the experimental subjects actually recognized themselves in their reflection. The problem arose because the mirror recognition had emerged as a stand-in for the previously dominant demarcator between human and nonhuman animals: language. In chapter 1, we see how a tradition of baby diaries emerged in the attempt to find a secure evidentiary basis for debates about the origin of language and thus human specificity. But in his baby diary, first composed around 1840 but published over thirty years later, Darwin argued that human language was closer to animal communication than had previously been supposed. Researchers were left scrambling for alternative demarcators, and they found one in the mirror, which had previously had a recurrent if marginal role in the baby-diary tradition.

In putting aside language, however, and thus focusing on nonlinguistic creatures, infants and nonhuman animals, psychologists like Wilhelm Preyer denied themselves one of the most powerful tools for understanding the behavior of their subjects. Without being able to ask them what they were experiencing, mirror researchers sought new ways of determining what their subject's reaction to the mirror meant. As we will see in chapter 2, the tension between the broader goals of the mirror recognition test and the difficulties of interpreting the results drove significant innovation in the practices involved. At first, in the 1880s and 1890s, this opened space for a range of women to engage in academic psychological research. But it also provided a neglected yet driving problematic for some of the canonical figures in the field throughout the first decades of the twentieth century.

The central tension of the mirror test was articulated most pointedly by the cybernetician Grey Walter (chapter 3). In the 1950s, he constructed a range of robotic "tortoises," which behaved in front of the mirror, he suggested, as if they recognized themselves. But whereas the other mirror researchers were unable to peer into the black boxes that were their subjects' minds, Walter could. This extra insight led him to mock those who thought, prematurely, that they had witnessed self-recognition. This tension continued to be the guiding problematic of the mirror tradition, until the introduction of the mark test, which was developed simultaneously by Beulah Amsterdam and Gordon Gallup around 1968 (chapter 4). Though it remained contested, the mark test provided a relatively secure means of determining self-recognition, by seeing what happened when the subject saw in the mirror a mark on their body that was not otherwise visible.

In solving the problem of how one could determine *whether* the subject could recognize its reflection, the mark test opened up space for another question: What does this recognition *mean*? This shift in questioning is exemplified by the career of French psychologist René Zazzo, who I discuss in the interlude. Upon integrating the mark test into his work, Zazzo started to tease apart the different stages of recognition, paying particular attention to the fact that the mirror image was an optical illusion. In part 2, "Misidentifications" I show how the illusory aspect of the image figured into a broad rethinking of the test. As we will see in chapter 5, already in the 1930s, Lacan had foreshadowed this shift in his famous "mirror stage," where a baby misrecognized its reflection, taking the unity of the mirror image as a sign that it too was psychologically one. He argued that the mirror image can be strange and alienating, shaping our views of ourselves as much as it reflects them.

In the 1970s and beyond, a similar insight motivated mirror researchers in a range of fields, who though they remained unaware of each other developed a strikingly similar set of claims. In chapter 6, I examine the work of Edmund Carpenter, who thought he had discovered a "mirror-naive" society among the Biami of Papua New Guinea. The experience led him to meditate on the deleterious effects of mirrors, disrupting previously held beliefs about the nature of media that he had drawn from Marshall McLuhan. For anorexia researchers in the 1980s and 1990s (chapter 7), the mirror represented both a point of access to the distorted sense of the body that they thought caused the disease, and a means to change it. Finally, in chapter 8 I examine a parallel development in the neurosciences, with the discovery of so-called mirror neurons. These mirrors were not physical objects; they had been internalized as crucial components of the brain. But in allowing an identification across difference, between self and other, mirror neurons resurrected and reformed many of the key questions of the mirror tradition, including, sociability, language, and human-animal difference.

PART I

Identifications

1

My Child in the Mirror

THE RISE OF THE MIRROR
SELF-RECOGNITION TEST

13th May [1840]. Three or four days ago smiled at himself in glass—how does he know his reflection is that of human being? That He smiles with this idea, I feel pretty sure—Smiled at my image, & seemed surprised at my voice coming from behind him, my image being in front.

. . .

10th [June].—When looking at mirror, was aware that the image of person behind was not real, & therefore, when any odd motion of face was made, turned round to look at the person behind.

. . .

Septemb. 23. When one says to him, "where is Doddy?". turns & looks for himself in looking-glass.—

. . .

Dec. 30th kissed himself in the glass & pressed his face against his image very like Ouran Outang [orangutan].
                    —CHARLES DARWIN, DIARY OF AN INFANT

This might seem a rather commonplace account of a child's behavior in front of the mirror.[1] Nothing that happens is particularly surprising. But the baby observed was not any child: Doddy was the pet name of William Erasmus, born on December 27, 1839, and the person he was looking back to in the mirror was his father, Charles Darwin (fig. 1.1).[2]

FIGURE 1.1. Charles Darwin and his first-born son William Erasmus ("Doddy") in 1942. Source: Wikimedia Commons.

Darwin's account of his son's behavior, a rare early foray into the realm of psychology, first appeared in a diary he kept between 1839 and 1856, detailing Doddy's development (to which Darwin added observations from his later-born children at the same developmental stage). This was a pivotal time of Darwin's intellectual trajectory, three years after returning from his voyage on the *Beagle*, and at a moment when he was working through the theory of natural selection, while studying the question of expression, especially emotion, in animals and children. The careful observation of his own children was part of Darwin's larger naturalist project, what he termed the "Natural History of Babies."[3] In fact, his baby diary is best seen as another of his scientific notebooks where he developed his main theories.[4]

Mirrors occupy a curious place in the diary. As the excerpt above suggests, Darwin developed an interest for them belatedly, considering Doddy's first response noteworthy only a few days after it happened. Subsequently, however, Darwin returned to the mirror behavior of his children insistently. Reflections, whether in a looking glass or in other surfaces, appear in over a fifth of all Darwin's observations from the first noted behavior at four and a half months to a little over one year old. The relevance of such encounters is not clear from the original diary, but Darwin picked them up at the other end of his career, when he returned to his observations while writing his 1872 book *The Expression of the Emotions in Man and Animals* and a short article for the journal *Mind*, "A Biographical Sketch of an Infant" (1877).[5]

In the latter, Darwin treated mirror behavior within (indeed as the primary object of) a section dedicated to the "association of ideas" and "reason," which as we will see had become central to arguments about human distinctiveness. The comparison of the child's mirror behavior to an orangutan in the quote above didn't make it into the 1877 article. In its place, Darwin added a new animal reference, focusing on the emotional response: While Doddy smiled at the mirror image, "the higher apes which I tried with a small looking-glass behaved differently; they placed their hands behind the glass, and in doing so showed their sense, but far from taking pleasure in looking at themselves they got angry and would look no more."[6] This was the only moment in the article where Darwin asserted a qualitative rather than simply a quantitative difference between human and animal behavior.

Darwin did not offer any extended analysis of Doddy's emotional response. We do not learn what we should take from the posited contrast with the anger elicited in animals, or what the mirror behavior said about the infant's cognitive development. Importantly for us, Darwin made no attempt to relate it to self-recognition. In his description of other mirror behaviors, he was most interested in whether Doddy understood the reflection as an image (and so whether he was surprised or not to hear his father's voice coming from elsewhere, or whether he associated his name with it).[7] It is perhaps due to the indeterminacy of its conclusions that Darwin did not consider his 1877 article to be particularly important, and later expressed his incredulity at its fate.[8] For it helped unleash a new and international tradition of mirror testing, which located in the child's response to its reflection the key to human specificity.

The indeterminacy and hesitancy are also fitting for a text that would launch the mirror-test tradition. The mirror test gained traction as a key demarcator between humans and other animals less on its own merits than as a response to the failure of another demarcator, one that had, up until that point, been closely associated with those two crucial categories in Darwin's analysis, the "association of ideas" and "reason," but which had been decisively challenged by Darwin's work. In the century leading up to Darwin's publication, human language was nearly universally embraced as the locus of human difference, and the means by which culture emerged out of nature. To understand why the mirror test assumed such an outsized importance then, we first have to trace the fate of this earlier demarcator, one that had become closely bound up with the practice of documenting a child's development in the first years of life.

Baby Diaries and the Question of Language

In the late eighteenth century, it seemed to many that the capacity for articulate language separated humans from all other animals. Such was the perceived gulf between human language and the communicative means of animals that the theologian, statistician, and permanent member of the Prussian Academy of Sciences Johann Peter Süssmilch argued that it could have only a divine origin. His 1766 *Versuch eines Beweises dass die erste Sprache ihren Ursprung nicht vom Menschen, sondern allein vom Schöpfer erhalten habe* found a wide readership, and in turn it led the Prussian Academy of Sciences in 1770 to organize an essay competition dedicated to the question.[9] As they put it, "Were humans, reliant only on their natural ability [*Naturfähigkeit*], able to invent language by themselves? And how did they do it?"[10]

The fifty-ducat prize went to Johann Gottfried Herder, at the time a young clergyman on the cusp of literary fame. In his essay, Herder agreed with Süssmilch about the uniqueness of human language—"the real differentia of our species from without, as reason is from within."[11] It pointed to the human disposition of *Besonnenheit*, which translates to something like thoughtfulness. *Besonnenheit* showed itself in the act of distinguishing one perception from another by attaching to it a sign, a "characteristic mark" (*Merkmal*). For instance, humans came to attach the *Merkmal* of bleating to a sheep (the "bleating one").[12] Nevertheless, Herder challenged the theologian by giving a purely naturalistic account of language's origins. Herder considered that language could emerge through a form of imitation, and he placed great emphasis on onomatopoeia. On this sensuous, "poetic" basis, the higher "heavenly, spiritual concepts" could be constructed.[13] In later writings, Herder expanded his claims about language into a broader cultural-historical analysis. Language to Herder was the key to the larger cultural experience of humankind. He expressed this idea perhaps most grandiosely in 1785: "No cities have been erected by the lyre of Amphion, no magic wand has converted deserts into gardens: but language, the grand assistant of man, has done these."[14]

Herder's was the most famous submission to the competition that year, but another entrant will hold our attention here. Dietrich Tiedemann, a philosopher and professor of Latin and Greek at the Carolinum in Kassel, had submitted a treatise with a similar title to Herder's: *Versuch einer Erklärung des Ursprunges der Sprache*, which received an encouraging "très bon" from the jury.[15] The positive, if not effusive, response is fitting. Like Herder, Tiedemann argued for the absolute distinctiveness of human language, while attempting to give it a natural origin.[16] But Tiedemann sought to build these arguments on a less speculative ground. For instance, in arguing for human uniqueness, he pointed not to the cognitive faculty of *Besonnenheit*, but to the fact (as he saw it) that animals lacked the physiological ability to produce articulated sounds, a claim that lent itself to empirical confirmation in the present.[17] Tiedemann's approach was an attempt to push back against a common weakness in the "origin of language" literature. It focused on a moment, the transition from nature to culture, that was deep in the past and inaccessible to rigorous study. Reconstructing a moment lost to time and history, scholars often had no more solid foundation than the thought experiment. Tiedemann's grappling with this problem explains why, when he welcomed his first child, Friedrich, into the world on August 23, 1781, just over a decade after the essay competition, he sat down to compose a baby diary.

In writing a baby diary Tiedemann was participating in a Europe-wide tradition, where observers, mostly parents, went into the nursery to note down their offspring's development.[18] The reasons for this varied: some tried to give more empirical grounding to the existing psychologies that were either philosophical (such as Johann Nicolas Tetens's *Philosophische Versuche über die menschliche Natur und ihre Entwicklung*) or fictional (such as Jean-Jacques Rousseau's *Émile*).[19] Others, such as the German educator Joachim Heinrich Campe, pursued more practical goals, gathering data for educational reform.[20] Yet others used the baby diary as a way to justify breaking with existing educational systems and carving out space for their own approach to parenthood.[21]

Tiedemann's ambitions were even more vaunted: he used his diary to help him discern the place of humanity in creation. Chronicling Friedrich's development from the moment of his birth to three years of age, Tiedemann thought he would be able to examine on a local and individual scale the social and historical developments he had tracked in the essay competition. Here the emergence of distinctive human traits could be studied up close in real time.[22] Tiedemann's *Beobachtungen über die Entwicklung der Seelenfähigkeiten bei Kindern* (Observations on the development of the power of the soul in children) appeared in 1787.[23]

While Tiedemann in his competition essay traced human-animal differences back to their distinct capacities for making sounds, in his *Beobachtungen*, the more secure evidentiary basis of the baby diary gave him license to give a more expansive answer, ironically moving closer to Herder's view. Tiedemann now located the difference between animals and humans in what he called the "higher powers of the soul" (*höhere Seelenkräfte*), the human capacity for judgment and comparison (*Urteil* and *Vergleichen*). The first signs of this difference became visible at four months when "children practice for quite a while by producing inarticulate tones and simple articulations."[24] Such practice helped them "to gain clear ideas [*Vorstellungen*] of [the sounds and articulations] before they begin also to repeat [*nachsprechen*] them."[25] The constant attempt to use comparisons to work out ever finer differences both in his own speech and in the signs around the child formed the basis of language, such that at eight months "when one asked him, where is this or that of the everyday objects around him, he pointed at them with his finger. He therefore has not only a clear sense of those objects, but also of the articulated sounds, and knew moreover, that these sounds denoted these objects or impressions."[26] This, for Tiedemann, was the decisive step. Friedrich had mastered the "most difficult of all associations of ideas, which the animal can only achieve in rare cases, with great effort, and never on its own, the association between an idea [*Vorstellung*] and its sign, the word."[27]

After Tiedemann, when baby diarists focused on animal-human difference, language again came to the fore. It appears in *Progressive Education* from 1828 by Albertine Necker de Saussure, daughter-in-law of Louis XVI's finance minister and cousin by marriage to Madame de Staël. And it had a privileged position in the physician Berthold Sigismund's *Kind und Welt* of 1856.[28] It was against this background that the French man of letters Hippolyte Taine (1828–93) set about examining the first years of his daughter Geneviève's development, an account that was published in the *Revue philosophique de la France et de l'étranger* in 1876.[29]

Taine mobilized the baby diary in order to contribute to an ongoing debate over evolution. In a set of lectures from 1873, the Oxford Sanskrit scholar Max

Müller had argued that the human language was decisive evidence that we could not be descended from the apes.[30] In human languages, for Müller, words referred to "general concepts" and could be applied to numerous distinct individuals.[31] This "rational" language could be identified, Müller argued, "among the lowest savages," while no trace of it could be found "even amongst the most advanced of catarrhine [i.e., Eurasian and African] apes."[32] Müller concluded from this that humans must possess a separate faculty of the mind, reason, that was denied to other animals. Or as he put it in a 1861 lecture, human language "is our Rubicon, and no brute will dare to cross it."[33] Through such linguistic reasoning, Müller concluded that, no matter how compelling the evidence Darwin drew from biology, "man cannot be the descendant of some lower animal, because no animal except man possesses the faculty, or the faintest germs of the faculty, of abstracting and generalizing, and therefore no animal, except man, could ever have developed what we mean by language."[34]

This did not mean that Müller was unwilling to explain the emergence of the rational out of emotional language. Müller suggested that humans might once have been bereft of rational language, before the onset of what he called the "radical period." Whether such a language arose as a "slow growth, or . . . instantaneous evolution," Müller declined to speculate.[35] Nevertheless, he spent much of his final lecture trying to understand how the transition might have occurred. He argued that rational language found its origins in the emotional, words that were "interjections, such as pooh, and imitations such as bow-wow."[36] The transition was facilitated by the ways in which humans produced an "infinite variety of imitations, many of which it would be almost impossible to recognize or understand, without traditional or social help," and here Müller referred to the confounding variety of onomatopoeia in different languages: the German cock says "*kikeriki*," the Chinese "*kiao*" and the Mongolian "*dchor.*"[37] These words then became "rational," when humans used them to denote larger groups of individuals: parents rather than just the singular mother or father; or large groups of animals that included more than one species.[38]

Taine's article has most often been assumed to be a refutation of Müller's argument.[39] But that is clearly not the case, at least not immediately or simply. He had written it in two sections, the first tracking Geneviève's first months and the second composed primarily of quotes from Müller's lectures focusing on broader linguistic questions. Taine opened that second section by noting his broad agreement with the Oxford scholar: His baby diary provided conclusions on the level of psychology that were "in all essential points the conclusions that M. Müller draws on the basis of philology."[40] Indeed Taine's parallel between the baby's development and that of the human species relied on Müller's account of the latter, the analysis of the dawn of the "radical period" in his final lecture. Babies started with the type of emotional language Müller identified as common to humans and animals alike (Geneviève's "twittering"). But they exceeded other animals in the variety of expressive "delicacy." And this, according to Taine, was the "source in him [man] of language and of general ideas."[41]

The observation of his child, however, allowed Taine to move the argument in a slightly different direction, a direction that is not noted as a divergence from Müller, but which fundamentally shifts the terms of the debate. Early on, Geneviève began a "spontaneous apprenticeship," relying on reflex action and some imitation, to form a range of different sounds, starting with vowels and slowly adding consonants. This resulted in a "very distinct twittering," which Taine compared to that of a bird, full of varying sounds, but lacking meaning. Through such processes, the child had forged for itself the "materials of language" by the age of twelve months. At this stage, Taine noted, Geneviève did depart from animals in the "delicacy and abundance of her expressive intonations."[42] But, he was clear, she had not breached the peculiarly human.[43]

Nevertheless, her capacity for more varied and subtle articulation laid the groundwork for the child to leave animal language behind, and start attaching generalized meaning to the sounds she uttered. At twelve months Geneviève pronounced her first real word: "*bébé*" (initially applied to any picture in the room, "*something variegated in a shining frame*") to which were quickly added, "*papa, mama, tété* (nurse), *oua-oua* (dog)" and others.[44] As the idiosyncratic use of the word "bébé" suggests, Geneviève's language was not simply learned from others but arose, Taine thought, out of an autonomous capacity for language: "It attaches to them [words] ideas that we do not expect and spontaneously generalizes outside and beyond our *cadres*."[45] The crucial point, which was the "distinctive trait of man," was that "two successive impressions, though very unlike, yet leave a common residue which is a distinct impression, solicitation, impulse, of which the final effect is some expression invented or suggested, that is to say, some gesture, cry, articulation, name."[46]

The slow and laborious development suggested to Taine that, pace Müller, the development could be traced to "a difference in degrees," related to "a greater development and a finer structure of the brain," and not the possession of a distinctive faculty.[47] As Taine summarized, "Therefore the monkey is on the same scale as the human but many levels below, without example or education ever being able to let it [the monkey] rise up to the level of the Australian, the lowest of humans."[48] When Taine finished his article by relating his child's first forays into language to that of so-called primitive societies, he added a line that implicitly challenged the main thrust of Müller's argument, that of denying the truth of evolution. The stepwise development of language in the child, Taine thought, was comparable to the way the "human embryo presents in a passing state the physical characteristics that are found in a fixed state in the classes of inferior animals."[49]

Taine's subtle but consequential divergence from Müller was accentuated when his baby diary was translated for the British journal *Mind* in 1877. The translator rendered only the first part in English, leaving the more positive account of Müller in the second part on the cutting room floor. It is perhaps for this reason that Darwin could find the article so amenable to his ideas, and why it might have led him to return to the baby diary he had written over a generation earlier. In his 1877 *Mind* article, Darwin stated that Taine's publication had "led me to look over a diary" he had kept of William, and bring out ideas that had not found outlet in his book *The Expression of the Emotions in Man and Animals*.[50] Thus in a strange inversion, Müller's attack on evolutionary principles was, through a two-way process of translation through French, and from linguistics through psychology to biology, transformed into a confirmation of Darwin's theories.

Darwin mobilized Taine's account as part of his broader argument against Müller to show the strong lines of continuity between human and animal language. Doddy, like Taine's child, invented his own word for food ("mum"). When saying the word, Doddy gave it "a most strongly marked interrogatory sound," which indicated for Darwin the persistence of instinctive elements: it was "analogous to cry for food of nestling-birds, which certainly is instinctive & peculiar to that time of life." Darwin concluded from this that, "before man used articulate language, he uttered notes in a true musical scale as does the anthropoid ape Hylobates."[51]

But it is important to note that Darwin had a more generous understanding of animal capabilities than Taine. Challenging Taine's identification of the first association of ideas at one year, Darwin put it at seven months, because at that stage the child had shown an ability to understand its parents' speech, even if Darwin agreed that the first word was spoken only around the first birthday.[52] Correlatively he identified a form of association in animals too, who could "easily learn to understand spoken words."[53] As he had written in *The Descent of Man*, "That which distinguishes man from the lower animals is not the understanding of articulate sounds for . . . dogs understand many words and sentences. In this respect they are at the same stage of development as infants between the ages of 10 and 12 months who understand many words and short sentences but cannot yet utter a single word."[54] The difference in powers was one of extent rather than of substance: "The lower animals differ from man solely in his almost infinitely larger power of associating together the most diversified sounds and ideas; and this obviously depends on the high development of his mental powers."[55] Or as he put it in the *Mind* article: "The facility with which associated ideas due to instruction and others spontaneously arising were acquired, seemed to me by far the most strongly marked of all the distinctions between the mind of an infant and that of the cleverest full-grown dog that I have ever known."[56] The combined effects of these moves was not to challenge simply Müller's claims that humans could not have descended from the animals, but also Taine's claims that human language marked a stepwise improvement over the animal.

Despite the ways in which Darwin's *Mind* article seemed to undercut the difference whose exploration had spurred on the baby-diary tradition, it gave that tradition a new lease on life. In *Mind* alone, multiple articles on the topic were published in the years following Darwin's, including accounts by the psychologists Frederick Pollock (1878) and James Sully (1880), and the physician Francis Henry Champneys (1881). Elsewhere, Darwin's article set off a lively and expansive culture of baby-diary keeping. We see the publication of at least twenty-eight book-length and thirty-two shorter studies between the 1880s and the 1930s, written by both men and women.

Darwin's account, however, had shifted the terms of the debate. Take the reception of Tiedemann's diary. Initially it appeared in the low-circulation *Hessische Beiträge zur Gelehrsamkeit und Kunst* in 1787, and it remained in relative obscurity for much of the next hundred years.[57] But in 1881, the French psychologist Bernard Perez edited, translated, and commented on it in French, a compilation that was in turn translated into English in 1890, seven years before the second German edition of the work.[58] Perez's edited version omitted what he saw as the "banal and superfluous," in order to focus on the "most substantial" of Tiedemann's findings, while placing them alongside his own discoveries as well of those of Taine and Darwin.[59] And on the question of language, Perez cleaves to the latter, explicitly contradicting Tiedemann's claims that the child left nonhuman animals behind at the eighth month, when it could achieve "association between sign and thing," something that "the animal can attain only rarely, with difficulty, and never by itself." For Perez, "this is a clear error. Between a child and a dog that associates the idea of *sugar* or of *meat* to the words that express these things, I don't see any difference from the mental point of view."[60] In his 1882 follow-up to the Tiedemann commentary, *La psychologie de l'enfant: Les trois premières années*, Perez doubled down on his claims about the commonalities between language in humans and animals.[61] After Darwin, it was hard to maintain that language marked the boundary between humans and nonhuman animals.[62]

Self-Consciousness and Mirrors: From Darwin to Preyer

Which brings us back to mirrors. Mirrors had appeared in baby diaries before Darwin, though they were rarely prominent and mostly subordinated to other concerns. The German deacon Immanuel David Mauchart in his 1798 diary is a case in point.[63] When his daughter Charlotte ("Lottchen") at eleven months was given a hand mirror and asked "Where is Lottchen?" "she looked behind the mirror, and when she did not find it [the image], she put the mirror very closely to her face to get closer to the image." For Mauchart this demonstrated that the child possessed "ideas of proximity and distance."[64] The mother featured in the educator Emma Willard's appendix to her translation of Necker de Saussure seemed most preoccupied with the question as to whether the mirror demonstrated the ability to identify an illusion.[65] We even get a foreshadowing of Darwin's discussion of an emotional distinction between animals and humans in front of the mirror in Sigismund's *Kind und Welt*, but Sigismund was primarily concerned with what the behavior told us about the child's sociability. For Sigismund, baby Geneviève's smiling at her mirror image (his son A. did so at twenty-seven weeks) indicated the "first lively expression of the impulse for sociability [*Geselligkeitstrieb*] that also prevails powerfully in many young and not a few grown animals."[66] Animals, however, rarely showed joy at the sight of their mirror image: "The monkey grinned at it, I saw dogs attacking it while barking, and ringed turtle doves cooing and bowing in front of it."[67]

The mirror was often discussed in the context of language. For Tiedemann, as we have seen, humans were distinct through their possession of "higher powers of the soul" (*höhere Seelenkräfte*), which consisted of the coupled functions of "judgment" (*Urteil*) and "comparison" (*Vergleichung*).[68] The presence of these higher powers became evident in various ways in addition to the "meaningful pronunciation of words," such as in memory or in the child's response to the mirror.[69] All three required discernment between similar items, for example, between similar sounds of words with different meanings;

between a past sensory impression and a current one (he gives the example of his preverbal boy who, having been to a specific place before, pointed toward it from a distance thus showing his recognition); or between similar visual images of people (the child's own reflection in the mirror and the appearance of others).[70] But while the mirror demonstrated higher powers of the soul, which was the focus of Tiedemann's research, it was not the only test, nor did it capture the first time when those higher powers became manifest. The "meaningful pronunciation of words" occurred first, at eight months, in contrast to seventeen months for the two other behaviors.[71] Language, therefore, kept its privileged place in Tiedemann's system.[72]

Given the way in which Darwin had seemingly sidelined the language demarcator, however, his vague assertion of human-animal difference with respect to the mirror came to assume an outsized importance. We can see this shift in Perez's treatment of Tiedemann. He challenged the notion that the "recognition" of the mirror image occurred only in the seventeenth month, arguing that it could be noted considerably earlier.[73] The claim fit into a broader argument that Perez made that cognitive developments did not correlate directly with language and thus that self-consciousness (which he located in both animals and humans), could be found before the child could say "I."[74]

The person who developed the claim most fully was the psychologist and physiologist Wilhelm Preyer in his *Die Seele des Kindes* (*The Mind of the Child*) of 1882, which would become the most influential baby diary of the late nineteenth century.[75] It went through nine separate editions before 1923.[76] Preyer argued that humans were distinct from nonhuman animals by their possession of the *Ichbegriff* (self-concept), and this could first be discerned in the child's response to the mirror.[77] Preyer was an ardent Darwinian. He had tried to get hold of the first edition of *On the Origin of Species* in 1859, though it was too soon out of print. When Preyer finally managed to track down a copy of the second edition in 1861, he was deeply impressed. He wrote the first doctoral thesis submitted to a German university that applied the Darwinian principle of survival of the fittest to a specific case (he argued that the principle could explain the extinction of the great auk, a flightless bird, in Iceland).[78] He subsequently became one of the great popularizers of Darwin's theories in Germany. In their correspondence—carried out over a decade—Preyer reported his advocacy to a grateful Darwin, who still felt in need of support.[79] After his habilitation, Preyer became the professor ordinarius for physiology at the University of Jena in 1869, the "Hochschule des Darwinismus" (university of Darwinism).[80]

Preyer's interest in Darwin was, however, mediated through the scientific context in Jena. Traditionally, psychology in the German-speaking lands had strong ties with philosophy. But Hegelian idealism, the dominant philosophy at the time, was widely perceived as having failed to keep up with the progress in the natural sciences, and for many, psychology needed to forge a new path for itself. For many, such as Hermann von Helmholtz and especially Wilhelm Wundt, this meant constructing and promoting a distinctive psychological method, different in particular from both philosophy and physiology. For instance, though they recognized that sensation depended on nervous function, they insisted that its study had to be independent of physiology; Wundt focused his attention on measuring the correlation of stimulus and reported sensation. In contrast, Preyer was skeptical of any psychology that ignored its physiological foundation. In his review of Wundt's foundational *Grundzüge der physiologischen Psychologie* (*Principles of Physiological Psychology*) of 1874, Preyer made this very clear. He expressed his view that one could either relate mental processes to their material correlates or reflect philosophically on the *Seelenleben* (activity of the mind). That is, either psychology could be philosophical (and thus unscientific) or it could instruct itself according to the most up-to-date physiology.[81]

When Preyer appealed to physiology, he placed himself in a longer tradition that used embryology to illustrate the evolutionary process, a tradition that was given its most popular articulation, "ontogeny recapitulates phylogeny," by the evolutionary zoologist Ernst Haeckel, Preyer's colleague at Jena.[82] In this view, embryology and evolution confirmed each other. Evolution helped embryology move from a descriptive to a genetic outlook, and embryology provided present-day evidence of Darwin's thesis: the development of embryos recapitulated the evolution of the species, moving through stages in which they resembled fish, then amphibians, and ultimately mammals. For Haeckel and others, ontogenetic development continued after birth.[83] When Preyer's first edition of *Die Seele des Kindes* appeared in 1882, Haeckel had already published his *Anthropogenie*, where he had tracked the "stepwise development of the child's mind" and deplored the absence of interest in the subject among psychologists.[84] Preyer too linked the study of the mental development of children to embryology; he published his *Specielle Physiologie des Embryo* in 1885, just three years after *Die Seele des Kindes*.[85]

It was in this context that Preyer encountered Darwin's *Mind* article, asking the latter for a copy in July 1877, the very month it appeared (as he pointed out, *Mind* was "a periodical scarcely known" in Germany).[86] When he received it, however, Preyer was not impressed. It was, he would later assert, "somewhat sketchy" and "does not have the same value that some, because of the author's name, have attributed it."[87] Its value lay rather in the "fact that a Darwin does not find it below his dignity to make all the doings of a child in the first years of its life . . . the object of his observations."[88] That is, Preyer thought it useful primarily for encouraging other studies, not least his own. For this, the timing was particularly fortuitous: Preyer's wife, Sophie, was at the time five months pregnant. She gave birth to a son, Axel, on November 23, 1877.

Over the next three years, Preyer observed the mental development of his child, a project that required unusual perseverance; he was present in the nursery "at least three times almost every day: in the morning, at midday, and in the evening."[89] Each day was to be recorded on a separate sheet, leaving the verso blank so that the notes could be organized later according to their content "through cutting out and gluing together in a new notebook."[90] Preyer was consumed with enthusiasm: "almost every day one could register some psychogenetic fact."[91]

Like many before him, Preyer was very concerned to distinguish between what was innate and what acquired, and in relation to this he discussed human-animal difference throughout the book. But in the wake of Darwin's work, Preyer refused to grant language its former importance as a demarcator. His argument drew on many of the claims from Darwin's article, not least the capacity of animals to understand (if not use) articulate language:

> Overall, one will want to rank the infant or the weaned young child in this stage of his cognitive development higher than a very intelligent animal, *but not because of his power of speech*. For the dog as well understands, in addition to hunting vocabulary, many words in its master's speech and guesses from his facial expressions and gestures the meaning of entire sentences, and even if it does not produce any articulate sounds, then the cockatoo, which learns all sounds of language, does so all the more. A child, who shows through facial expressions, gestures, and acts that he understands single words, who already imitates many words without understanding, but only speaks a few, and without understanding, outranks intellectually a smart and calculating and yet alalic [without speech] elephant or an Arabian horse intellectually not for this reason, but *because he forms more and more complicated concepts*.[92]

This did not mean that language was unimportant, for it could refine and make more precise (*präzisieren*) "primitive and unclear concepts" and "can develop even further through connecting the ideas to the circumstances in which the child lives, making higher and higher abstractions."[93] But it was secondary and subsequent to the powers of concept formation.

Most importantly, for Preyer animals lacked the crowning concept of man, that of the self, the *Ichbegriff*. This concept was discussed in the last substantial chapter of the book, marking the end point of early child development and Preyer's account of it. The *Ichbegriff* could not be tested with language. In line with Perez and his other pronouncements on the subject, Preyer did not think that the use of the personal pronoun *I* was very significant; it emerged after children had formed their *Ichbegriff*: "Many stubborn children have a strongly developed *Ichgefühl* [self-feeling] without referring to themselves other than with their name," which was the result often of parents referring to themselves as well as the child in the third person.[94]

The *Ichbegriff* was first revealed, rather, in the child's reactions to a mirror. At first, the child was indifferent to the mirror image. Around week seventeen, the child looked at his image "for the first time with unmistakable attention, with the same expression with which he fixates a strange face that he sees for the first time."[95] A few days later he smiled at it. In week twenty-four, the child turned around to Preyer, whom he saw in the mirror standing behind him and reached for the mirror image with his hand in week twenty-five. At week thirty-five the boy was puzzled when he sought to touch it and instead felt a "hard smooth surface."[96] His puzzlement turned into unease and rejection and when the child was presented with a mirror again, "he immediately turned away."[97] Preyer wrote about this moment: "Here the *ungraspable* [*das Unbegreifliche*]—in a literal sense—was unsettling."[98] This type of engagement with the mirror image continued until in week sixty, when, "responding to the question 'Where is mama?' he pointed to the mirror image and then turned around to the mother while laughing." Preyer concluded that "now, after 14 months, original and image were reliably distinguished."[99]

The child was now ready for the next stage. In week sixty-seven, the child made grimaces in front of the mirror that made him laugh, and he started to show "signs of vanity" in week sixty-nine, such as dressing up by wrapping a lace or embroidered cloth around his shoulders as in a train, and at the seventeenth month started to make grimaces at its reflection.[100] Because he considered neither the grimaces nor the dressing up to be welcome behaviors, Preyer decided to terminate the mirror experiments (*Spiegelversuche*) at this point.[101] But taken together, Preyer thought, these responses showed the "transition from the *ego-less* [*ichlosen*] state of the newborn who cannot yet clearly see to the state of the developed self."[102]

Preyer was clear that here humans differed greatly from animals. The way animals responded to the mirror suggested that they saw other animals in the reflection, not themselves. Immediately following the chronicle of his child's mirror responses, Preyer presented his famous example of the Turkish duck:

> A couple of ducks from Turkey, which I saw once a day for several weeks, always kept apart from other ducks. When the female duck died, the male liked to swim toward a small strongly reflecting basement window and every day stayed before it for hours on end. . . . He perhaps thought that it was his lost companion.[103]

So too, the cat in Preyer's home, when confronted with the mirror, "must have taken the image for a second living cat, for when the setup of the mirror allowed, the cat walked behind and around it."[104]

Preyer did not comment on the higher apes, but others at the time pointed out that they did not seem to have a self-concept either. As the Frankfurt zoologist Maximilian Schmidt pointed out in 1878, the orangutan that he observed "saw and recognized in the mirror the onlookers present, for he fixated them for a while in the image and then looked around for them, as if he wanted to make sure they were really there."[105] Nonetheless, the animal did not seem to be able to identify with its own mirror image. We hear of the animal's horrified reaction upon first discovering the ape in the mirror, the raising of his hackles, the pushing forward of his lower lip, his backing away from the mirror. Later he sought to attack the image by spitting at it, hitting it with a wooden hammer, and throwing bread crumbs. From these observations, Schmidt drew the conclusion: "That he recognized himself in the mirror image was not verifiable, for he made no movements or grimaces at all, which would probably not have been absent if the meaning of the phenomenon had been clear to him."[106] This was also the experience of the zoologist Johann von Fischer of Gotha, studying the actions of a mandrill (a small African primate) who showed his behind to his mirror image. "He walked toward it [the mirror], came to a halt at some distance in front of it, laughed at the image, and turned around immediately to show the colored parts of his body."[107] Fischer did not further comment on this action in his article; but Darwin, discussing the publication a few months later, saw the action as clearly part of courtship behavior, which implied that the mandrill saw a potential mate in the mirror, rather than an image of himself.[108]

The Mirror in the Brain

The confidence with which Preyer read self-consciousness into the child's mirror behavior might surprise us. As we have seen, the mirror had rarely been related to self-consciousness before, and certainly not in Darwin's account. In part this was due to the nature of mirror recognition. Was identifying with the mirror image really a sign of self-consciousness? After all the mirror image was distinct from the self, not least spatially, and it was not clear that recognition of the image was akin to the type of acts usually associated with self-consciousness, such as reflection on one's acts. Preyer latched onto the *Ichbegriff* for the way it sidestepped this problem. The *Ichbegriff* was required *both* for consciousness of the self—it was that concept to which attributes gleaned from reflection on one's acts could be added—*and* for mirror recognition, where it was applied to the mirror image. These concerns also explain why Preyer placed great weight on the distinction between self and reflection. As he argued, the existence of the "developed self" was one that "consciously distinguished itself from the mirror image and from others and their mirror images."[109]

But even if mirror recognition could be seen as a form of self-consciousness, it was not clear that such recognition was demonstrated in the behaviors Preyer observed. Very few of the responses noted by Preyer or his predecessors unambiguously suggested that the child saw the image as a reflection of its own body. Preyer admitted as much, stating that he could "follow all these advances in detail only with great effort."[110] In *Die geistige Entwickelung in der ersten Kindheit*, the popular version of *Die Seele des Kindes*, Preyer offered instructions in the form of guiding questions to parents keeping their own baby diaries.[111] In the section "The Mirror Image," the moment of self-recognition is notably absent. What Preyer focuses on instead are external behaviors: "When does [the child] attempt to grasp behind the mirror? When does he try to kiss his mirror image?"[112] Not only was the moment of self-recognition hard to discern; Preyer did not think it could be pinpointed exactly. Rather, he declared, it was only when the mirror behaviors were taken together that they "form something like converging lines, which culminate in the perfect feeling of unity of the personality and its demarcation from the external world."[113] Even then Preyer settled on the mere assertion of convergence, without relating it to the behaviors out of which it was composed.

If there was no clear moment of self-recognition, nor any clear articulation of how the documented behaviors manifested a gradually coalescing self, why was Preyer so sure that they proved the existence of the *Ichbegriff*? To understand his confidence, we need to turn to another aspect of his work: neurology, which was closely associated with his engagement with Haeckel's physiology. Preyer was not undertaking neurological research at this point. Rather, he was drawing on existing neurological knowledge in order to understand the behavior of the child before the mirror. We can go further and suggest that Preyer deployed a neurological hermeneutic, which underwrote his reading of the child's behavior and especially the mirror encounter while giving that reading a scientific sheen.

Preyer framed his analysis in *Die Seele des Kindes* with a discussion of neurological association. Drawing on theories about sensory development proposed by Douglas Alexander Spalding (an English biologist who, earlier in the century, had become notorious for his experiments on newborn chicks) and some of his own, he discussed the sight of newborn animals, their hearing, sense of smell, and so on.[114] Preyer argued from this that even though newborn animals had a greater perfection of their sensory system at birth, humans were born with a more advanced ability to form associations. The "more or less often repeated use of specific association pathways in the cerebrospinal system" enabled humans to acquire a much greater number of sensory associations with movements than could the animals and thereby "to learn other usages."[115] This consequently allowed humans to build concepts, which ultimately made humans different from animals.[116]

Preyer then returned to this associationist neurology immediately after his discussion of mirror behavior.

> The faculty (the ability, the predisposition, the potential function) to form concepts is innate, and some of the first concepts are hereditary. New (not hereditary) concepts only develop after new impressions, that is, experiences, which make connections with the primitive concepts, through new fiber systems in the brain, and this begins before the acquisition of language.[117]

These neurological assumptions informed his reading of the mirror behavior. Preyer was interested first and foremost in the way the child draws on the associative powers of the nervous system to unify its body. In the newborn, different parts of the nervous system (most importantly the spinal cord and brain) were still independent from each other: "The cortical self [*Rinden-Ich*] is different from the spinal cord–self."[118] This explained why at an early stage the child took its limbs "as if they were something foreign."[119] As the child grew older, the "various sensory areas" became increasingly connected through "intercentral fibers," forged through experience, whose combination marked for the child its own body: the "frequent coincidence of disparate sensory impressions during tasting-touching, seeing-touching, seeing-hearing, seeing-smelling, tasting-smelling, hearing-touching."[120] That explained why after about the sixty-second week, the child's interest in its limbs abated, and Preyer concluded that this signified "the definite separation of object and subject in the child's intellect."[121] These processes illustrated the development of a "obscure feeling of self" (*dunkles Ichgefühl*), which would later, "through further abstractions," turn into the fuller and more clearly defined *Ichbegriff*: "'The advances of the mind while *looking at the mirror image* confirm this conclusion from the observations above."[122] Or as Preyer translated it into neurological language, they formed the basis for the "total idea of the self" (*Gesammtvorstellung des Ich*), which was produced by the simultaneous excitation (*Erregung*) of the "cerebral centers of all sensory nerves."[123] Preyer thought that the psychological development of the child culminated in the *Ichbegriff* and that he could see that process unfold in front of the mirror, because that was the telos of the process of nervous association building in the brain.

Conclusion

Darwin's 1877 article was more important for the way it delegitimized particular arguments than for the way it proposed new ones. As we have seen, baby diarists were engaged in a paradoxical project of explaining the origin of a discontinuity, for Tiedemann and others literally a primal leap, *Ur-sprung*, from nature to culture. They thought they could both explain language and give it a history, as the foundation and precondition of reason, and determine it as a qualitative difference separating humans from the rest of the animal kingdom. In his article, Darwin sought to tell the same history, drawing even on the same analogy between the development of the species and the development of the child. But in casting this development as evolution, he argued that the history of language undercut any claims about discontinuity: given the way in which a child's babbling could be related to the twittering of birds and other animal sounds, human language could not be seen as qualitatively different from animal communication.

Darwin's argument was clearly convincing. After its publication, psychologists found it increasingly difficult to draw the border between humans and other animals using language. Nevertheless, and if only in a limited, suggestive way, Darwin's baby-diary article offered another marker. Doddy's smile, so different from the aggression displayed by other animals in a similar situation, gestured to developments in the child's mind that might set humans apart. That this occurred in front of a mirror, where the recognition of the image seemed so tantalizingly close to the recognition of the self, helped tie the test to a longer tradition that placed self-consciousness at the root of human cognitive difference.[124] The mirror test promised to be a new nonlinguistic shibboleth.

But however intuitive the self-consciousness reading of mirror behavior was, the structure of the mirror test made it difficult to ground it in any secure way. Doddy's smile occurred at four and a half months, long before anyone expected the child to have a self-concept. This would only become clearer as Preyer's ideas were picked up in the English-speaking world, where the prerogative to investigate child development was taken from the parent-scientist and invested in a newly developing scholarly discipline that looked beyond the single child. How would the self-concept fare in a scientific environment that favored easily repeatable tests, with unambiguous results?

2

"Not Suddenly, but by Degrees"

CHILD PSYCHOLOGY, GENDER, AND
THE AMBIGUITY OF THE MIRROR

FOR HER 1931–33 BOOK *The First Two Years*, the University of Minnesota child psychologist Mary Shirley engineered a "cooperation between home and laboratory" to map the developing capabilities of small children. She evaluated twenty-five babies every other week according to a set of standardized tests.[1] These included "choice tests," which established the child's preferences for specific colors or specific toys; "manipulation tests," such as opening and closing a box, stacking blocks, or throwing a ball; and "picture and odor tests." In the latter, she recorded a range of "reactions to the mirror" (fig. 2.1).

The shift from Preyer could not be more stark. First, this was emphatically not a baby diary. The analysis wasn't tailored to a single child, noting whatever behaviors the child exhibited at any one time. Second, and in consequence, the major pieces of evidence were not dates, the time at which a development took place. Instead we have a series of percentages, indicating the number of children who had demonstrated the behavior at a number of set times, thus moving away from any clear sense of a particular trajectory or order. And finally, we seem to have a dramatic shift in the status of the mirror. Where Preyer had taken the mirror test to be the pinnacle of his study, demonstrating the emergence of the self-concept, the defining feature of humanity, Shirley saw it as but one out of a whole battery of tests, where one examined a range of discreet but relatively unremarkable behaviors; mirror self-recognition was nowhere to be found.

Shirley's list of reactions is representative of a broader transformation of the mirror test in the first decades of the twentieth century, especially as it developed in the United States. In the early days of academic psychology in

TABLE XLIII
REACTIONS TO THE MIRROR

REACTION	PERCENTAGE REACTING		AVERAGE TIME OF REACTION IN SECONDS	
	42 Weeks	50 Weeks	42 Weeks	50 Weeks
Attention				
Passive attention.............	21.0	23.8	11.5	10.6
Look at examiner or audience...	26.3	14.3	6.6	22.0
Look around................	0.0	0.0	0.0	0.0
Manipulation of mirror				
No reaction.................	0.0	4.7	0.0	5.0
Reach, lift, and take..........	68.5	90.5	13.0	8.9
Pat, scratch, or rub on head....	31.5	33.3	11.5	11.0
Passively hold or drop........	68.5	62.0	16.0	16.9
Look at it..................	36.8	19.0	13.4	15.8
Drop, as in a game...........	0.0	19.0	0.0	15.5
Turn mirror to look at back....	63.1	62.0	10.8	13.8
Look in it and point at image...	52.5	52.5	13.7	15.7
Chew or lick................	84.2	57.3	27.7	24.5
Try to escape...............	15.7	9.5	14.3	22.5
Bring examiner into situation...	5.2	14.3	5.0	13.0
Kiss image.................	10.5	23.8	12.5	13.0
Average point score..........	2.2	2.9	...	...

FIGURE 2.1. "Reactions to the Mirror." Mary Shirley, *The First Two Years: A Study of Twenty-Five Babies* (Minneapolis: University of Minnesota Press, 1931–33), 2:263.

America, researchers sought to combine Preyer's vaunted goals with a new sense of methodological rigor, drawn from another German source: Wilhelm Wundt. They prioritized the collating of data from multiple children and, downplaying neurology, sought to sidestep the interpretative impasses that had led Preyer to lean heavily on schemas of nervous development. As we will see, this situation briefly opened up a space for networks of women scientists across the United States who mobilized their domestic responsibilities in the service of their professional and academic goals.[2]

Though these networks did not survive the pernicious casting of the discipline as a masculine enterprise in the newly founded institutions of psychological research, they draw our attention to another constitutive strand of American psychology, rooted not in Wundt but in Preyer. For canonical figures like G. Stanley Hall and his child study movement, the problematic arising from Preyer's work drove their research, especially when it concerned the prelinguistic child. Like the women researchers before them, these psychologists responded to the mismatch between the developmental ambitions of the Preyer tradition and the data it was able to collect by moderating their claims and developing an experimental system that foregrounded above all iterability and consistency of interpretation. These concerns help explain the success of behaviorism in interwar psychology, as well as the changing status of the mirror as a tool for mental testing.

The Problem of Interpretation

As we have seen, whatever the intuitive plausibility of Preyer's self-concept reading, it was very difficult to find sure grounding for this interpretation in mirror behavior. What actually counted as evidence of self-recognition in a prelinguistic child? Even Preyer restricted himself to offering a sequence of behaviors in front of the mirror and did not identify a definitive shift. He was able to convince himself that this sequence demonstrated the emergence of the *Ichbegriff* only, if we remember, because he saw those behaviors as the outward manifestation of a nervous system slowly knitting itself together

These difficulties were heightened by the development of psychology in the English-speaking world, especially the United States, that tended to follow the inclinations of Wilhelm Wundt. Wundt sought the scientificity of psychology in its own rigorous methods and not in its relation to physiology (though he had borrowed his experimental style from that field). When visitors and students of Wundt brought the new psychology to the United States during the 1870s and beyond, they carried these ideals to the new psychological institutions established in the last decades of the century: psychology laboratories and departments at Columbia, Johns Hopkins, and Clark, the latter two newly founded universities; journals such as the *American Journal of Psychology*; and associations like the American Psychological Association (APA).[3]

To be sure, some psychologists, even among those interested in development, continued to lean on neurological theories. Take for instance the New York physician Elizabeth Stow Brown, who combined her own observations made at the New York Infant Asylum with the record from "intelligent mothers among [her] acquaintance" in 1890.[4] Her account of nervous associations preceded and informed her reading of the mirror experiments, which she took as a means of tracking multisensory association.[5] So too, Frederick Tracy, who in a revised version of the doctoral dissertation he wrote at Clark University in 1893 framed his account of child development by a presentation of neurological development.[6] This set the groundwork for him to cast the mirror as one of four kinds of "external evidences" that the child had become self-aware.[7]

Most others downplayed the reference to neurology. In Britain, while the child psychologist James Sully acknowledged that brain development was the underlying factor of mental development—a human infant's "unpreparedness for life" at birth, which was the most significant difference from animals, could be traced to the immaturity of its nervous system—he recognized that too little was known about these underlying processes to make them of use to the psychologist.[8] Rather than waiting for the scientists to provide a more secure neurological foundation, Sully argued that psychologists should proceed with their own observations. The University of Oklahoma–based developmental psychologist David R. Major voiced a similar criticism in 1906, taking aim at Preyer: "Preyer's interest, at first, was in the phenomena of physical development, and his work, *Die Seele des Kindes*, though rich in the data of child psychology, has—from the psychological point of view—the defects of a work written by one who was physiologist rather than psychologist."[9] Clearly, this was partly the territorial posturing of a psychologist asserting independence, but what is important here is that it was posed as a methodological shift: while not denying the role played by the nervous system, Major focused his attention at the level of behavioral description and psychology.[10]

Psychologists (and developmental psychologists in particular) no longer drew their scientific bona fides from an adherence to neurological materialism; they did so increasingly by their method, multiplying the number of test subjects, and controlling more carefully the conditions of experimental observation. In this, psychologists responded to what was increasingly seen as a significant weakness of the baby-diary approach. The baby diary had gained prominence as a simple-to-conduct domestic test, shaped by the intimacy and stability of the bourgeois family. But the idiosyncrasies of particular family lives or particular children made it difficult to build broader theories on this basis. As the baby-diary tradition expanded, the dangers of individual idiosyncrasy stood out ever more clearly. For instance, while Darwin had noted Doddy looking back to him after seeing his father's reflection at five and a half months, others such as Ernst and Gertrud Scupin, in their diary *Bubis erste Kindheit* (1907), first noted the same behavior at ten months, and Preyer another five months later.[11] So too on the question of associating the child's name with the mirror image: Some including Darwin noted it at eight and a half months, and it was on this basis that Bernard Perez, when translating Dietrich Tiedemann, challenged his seventeen-month dating of mirror "recognition."[12] But a later dating won out: The Scupins noticed it at twelve months, Sully not until the twenty-first month, and Major not until over half a year later (twenty-eight months).[13]

Preyer and others had already recognized this problem.[14] In his earlier work Preyer had emphasized the value of determining the sequence of development, which he thought could be best gleaned from a single well-studied example.[15] As he wrote at the time, "Even though one child develops quickly, another slowly, and the greatest individual differences occur even among children of the same parents; but the differences refer much more to the times and degrees than to the order and appearance of the specific moments of development."[16] And these, Preyer was certain, "are the same for all."[17] Even then, he suggested that the comparison of several accounts of baby development would help protect against unfortunate generalizations. As his work progressed, Preyer became more convinced of the necessity of a collective approach to data gathering. In *Die geistige Entwickelung in der ersten Kindheit*, Preyer noted that "a lot of work is yet to be done" and interest in childhood development needed to be "awakened in wider circles, and, where it already exists, [has to be] expanded," for the "scientific value" of these studies and the "practical benefit of the study of the child's mind for the highest and finest task that there is—*education*."[18] As we saw in the previous chapter, Preyer included practical instructions to would-be baby-diary researchers: He stipulated that one should keep one diary per child so as not to mix up findings, and strictly separate observational fact from interpretations. Only then could the baby diary "be used scientifically."[19] For similar reasons, James Sully wrote a note in *Mind* in 1893 soliciting parents and teachers to send him observations "during the first five or six years of life," to be arranged under thirteen headings (Attention and Observation; Memory; Imagination and Fancy; Reasoning; Language; Pleasure and Pain; Fear; Self-Feeling; Sympathy and Affection; Artistic Taste; Moral and Religious Feeling; Volition; Artistic Production).[20]

A Network of Mother-Scientists

Anxieties about the epistemological soundness of the individual baby diary briefly opened up a space for a range of women researchers in the United States, who were otherwise excluded from much scholarly activity.[21] Consider the educational reformer Emily Talbot, secretary of the Education Department of the American Social Science Association. In 1881, she sought to draw on the collective resources of America's parents, asserting the "importance of making some systematic effort to record the development of infant life."[22] Investigators had long been preoccupied with the study of animals, but although "the natural development of a single child is worth more than a Noah's Ark full of animals," only "little has been done in this study, at least little has been recorded."[23] That is why she proposed "a continued series of observations, in large numbers," which, she thought, "may reveal order in the variations of phenomena, and that some portion of the secret of the mental and physical development of infants may be discovered."[24]

With this in mind, Talbot published a questionnaire in various journals, primarily the *Journal of Social Science* but also in *Nature*, soliciting the help of mothers and fathers in observing their children in order to coordinate child observation on a mass scale (fig. 2.2).

The project was clearly rooted in the baby-diary tradition.[25] In an 1882 publication, *Papers on Infant Development*, which functioned as a second call for participants, Talbot included references to a range of works. She also included copies of the relevant literature, reprinting Taine and Darwin's *Mind* publications and a selection from Preyer's *Die Seele des Kindes* in translation, only a few months after its publication in German. For the participants in the study, these texts were to be "a guide to the manner of proceeding with the work of observation."[26]

It doesn't seem that Talbot ever published the results of her project, but her approach was picked up by others, including Annie Barus Howes, secretary of the Association of College Alumnae (ACA)—who encouraged mothers to "pluck up courage to pursue an investigation on our children's development"[27]—as well as the Society for the Study of Child Nature in New York City (organized by Mr. and Mrs. Felix Adler), the National Congress of Mothers, and women's clubs and parent-teacher associations across the country.[28] The most successful attempt to gather and compile multiple baby biographies, however, was undertaken by Milicent Shinn. As Christine von Oertzen details in her account of Shinn and her network, Shinn's career trajectory was both conditioned by, and an attempt to overcome, the position of college-educated American women. After completing her bachelor's degree at Berkeley, Shinn had been forced to return home to take care of her family in Niles, California, where they owned a fruit ranch. As the only daughter in a large household, she committed to the care of her parents and the education of her younger brother.[29] In this context, the baby diary offered her the opportunity to develop her scholarly credentials despite being restricted to the domestic sphere.

APPENDIX.

CIRCULAR.

The Department of Education has issued the following Circular and Register:

We have been made familiar with the habits of plants and animals from the careful investigations which have from time to time been published,—the intelligence of animals, even, coming in for a due share of attention. One author alone contributes a book of one thousand pages upon "Mind in the Lower Animals."

Recently some educators in this country have been quietly thinking that to study the natural development of a single child is worth more than a Noah's Ark full of animals. Little has been done in this study at least little has been recorded. It is certain that a great many mothers might contribute observations of their own child's life and development, that might be at some future time invaluable to the psychologist. In this belief the Education Department of the AMERICAN SOCIAL SCIENCE ASSOCIATION has issued the accompanying Register, and asks the parents of very young children to interest themselves in the subject,—

1. By recognizing the importance of the study of the youngest infants.

2. By observing the simplest manifestations of their life and movements.

3. By answering fully and carefully the questions asked in the Register.

4. By a careful record of the signs of development during the coming year, each observation to be verified, if possible, by other members of the family.

5. By interesting their friends in the subject and forwarding the results to the Secretary.

6. Above all, by perseverance and exactness in recording these observations.

From the records of many thousand observers in the next few years it is believed that important facts will be gathered of great value to the educator and to the psychologist.

FIRST SERIES.

REGISTER OF PHYSICAL AND MENTAL

Development of 'Give the Baby's' _____
                   full name

Name and occupation of the father? _____

Place and time of father's birth? _____

"   "   mother's " ? _____

"   "   baby's   " ? _____

Baby's weight at birth? _____ at 3 months? _____

"   6 months? _____ a 1 year? _____

Is baby strong and healthy, or otherwise? _____

At what age did the baby exhibit consciousness, and in what manner? _____

AT WHAT AGE DID THE BABY

smile? _____

recognize its mother? _____

notice its hand? _____

follow a light with its eyes? _____

hold up its head? _____

sit alone on the floor? _____

creep? _____

stand by a chair? _____

stand alone? _____

walk alone? _____

hold a plaything when put in its hand? _____

reach out and take a plaything? _____

appear to be right or left handed? _____

notice pain, as the prick of a pin? _____

show a like or dislike in taste? _____

appear sensible to sound? _____

notice the light of a window or turn toward it? _____

fear the heat from stove or grate? _____

speak, and what did it say? _____

HOW MANY WORDS COULD IT SAY

at 1 year? _____ at 18 months? _____ at 2 years? _____

Will the mother have the kindness to carefully answer as many as possible of these questions and return this circular, before July 15th, 1881, to   MRS. EMILY TALBOT,

Secretary of the Education Department of the American Social Science Association,

BOSTON, March 1, 1881.   66 MARLBOROUGH STREET, Boston, Mass.

In connection with the inquiry indicated above, the following letter from Dr. Preyer, of Prussia, addressed to Mrs. Talbot, will be found of interest.

FIGURE 2.2. First version of Talbot's register of infant development. Parents were encouraged to fill out the questionnaire and return it to her. *Journal of Social Science* 13 (April 1881): 189–91.

Shinn carefully observed and documented the development of her niece Ruth from the sixth day of Ruth's life up to the age of seven.[30]

In 1891, one year after Ruth's birth, Shinn launched the ACA California Child Study Section. She recruited ten recent college graduates who had young children (all born between 1889 and 1891) and held regular meetings to discuss findings and the latest research literature on infant development, invite experts to their meetings, and build personal as well as institutional connections. The result was a number of publications including the multivolume *Notes on the Development of a Child* (1893–1908). In 1898, Shinn was awarded her PhD in education for this work, through an alliance with the Graduate Seminar in Pedagogy, which she had helped arrange.[31] Shinn's Berkeley affiliation and degree facilitated her participation in intellectual exchange across the country; she presented her work at national conferences and corresponded with American and European leaders on the topic.

Like Talbot, Shinn saw her project as an extension of and improvement on Preyer's. Reading a translation of Preyer with her sister-in-law, she embraced his phylogeny-ontogeny argument. In a popular version of her account, *The Biography of a Baby* (1900), she noted that "one aspect of especial interest in genetic studies" was "the possible light we may get on the past of the human race."[32] Even before Darwin, observers had noted the "resemblances between babies and monkeys," and recent embryology, she thought, had only served to confirm this: "Each individual before birth passes in successive stages through the lower forms of life."[33] Because it was not possible to record "the passing over of man from brute to human," the ontogenic analogue in the growing child was a valuable object of study, promising to fill in the gaps in our phylogenetic knowledge.[34]

Nonetheless, Shinn worried about Preyer's "biographical method." By tracking the development of a single child, he failed to see beyond the peculiarities of his chosen subject. She thus proposed combining Preyer's method with the "comparative" or "statistical" approach, which had traditionally been used on school-age children. But in its usual form, this had failed to give insights into the "successive steps as they actually take place in one and the same child," and thus the order of mental development. Shinn thought that by collating multiple biographical records and then "compar[ing] [them] throughout with other observations," she would be able to make generalizable claims, while retaining an ability to track the path children took.[35]

These concerns help explain a significant shift in the reception of Preyer. For, though Preyer had placed the mirror test at the pinnacle of early child development, in Shinn's and Talbot's work it dropped out almost completely.[36] This had not been the plan. At the beginning, Talbot placed self-consciousness at the heart of the project. Almost certainly at Talbot's request, her daughter Marion translated Preyer's earlier "Psychogenesis" for the *Journal of Speculative Philosophy* in 1881, an article that made brief claims about the emergence of self-consciousness, but did not refer to the mirror.[37] In preparing the manuscript, Talbot senior had written to Preyer to inquire further about the topic: Could a child of eleven months "expressing its wishes and inducing the nurse to comply with them" be considered as exhibiting self-consciousness? Preyer answered in the negative, but he acknowledged that

> this is one of the most intricate questions to decide—*when* the child distinguishes its own body, head, hands, &c., from other objects, as belonging to himself. The first time a child says "I" and "me," in the correct sense, it may be considered to have passed the limit. The formation of ideas by associating impressions, as well as the formation of general ideas (*Begriffe*) by uniting similar qualities of different objects, is intellectual work done by the child long before it knows anything of its own individuality. It seems to be that self-consciousness does not arise suddenly, but by degrees, after many experiments have shown the difference between touching his own body and external objects with his little hand.[38]

In the same letter, Preyer announced the imminent publication of *Die Seele des Kindes*, which Talbot again passed on to her daughter. Marion Talbot translated the chapter on the mirror for the American journal *Education* in January 1882, immediately after the book's publication in Germany, and well before the complete translation of the book in 1888. She gave it the title "Notes on the Development of Self-Consciousness."[39] Talbot senior then reduced the chapter to its essentials for the 1882 volume *Papers on Infant Development*, removing the "speculative parts of the translation" on neurology, and preserving only the descriptive account of mirror behavior.[40]

While Talbot was interested in the question of self-consciousness and consequently the mirror test, she wasn't able to package it in a way that fit her questionnaire methodology. She wanted to gather together observations from across the country and in a variety of different situations, so the questionnaire necessarily asked questions that invited short, unambiguous answers. Talbot inquired after the precise age of milestone developments such as the first smile, when the baby first recognized its mother, held up its head, appeared sensible to sound, and so on.[41] These could then be collated and analyzed using statistical tools. But how could the mirror test, with all its complexity, fit this model? Talbot lacked an answer. Instead she included one, more open-ended, question: "At what age did the baby exhibit consciousness, and in what manner?"[42]

The reference to consciousness didn't help things. One mother wrote: "The question as to the earliest exhibition of consciousness, seems to me a little ambiguous. Conscious of hunger, of the difference between arms and the bed, certainly,—yet I doubt that being the meaning of the question."[43] In her response, Talbot addressed this problem, clarifying that it was "*self*-consciousness" that was at issue. In fact, she argued, it was one of the "objects of these studies," "to throw more light on the growth of self-consciousness in young children from the parents' point of view," again pointing to Preyer, alongside Perez, as models.[44] Nevertheless, and perhaps to ensure that the results would be comparable, Talbot did not materially change the second version of the questionnaire in 1882, and the ambiguous wording of the *consciousness* question remained. If even Preyer was unable to identify a single behavior that demonstrated the existence of self-consciousness, what hope was there of finding a standardized protocol that could be replicated in homes across the country? Mirror testing simply wasn't compatible with the new scientific ideals at work in the United States.

Shinn seems to have recognized as much. We see this when we turn to one of the few instances where a member of Shinn's network took notes on mirror behavior: Laura Swain Tilley's "Record of the Development of Two Baby Boys." Tilley referred to mirror behavior eight times in her record, noting the child's positive emotional response to the mirror at various moments (at the ages of one month, seven months, and ten months), his surprise at seeing a baby beside his mother in the glass (day 123), and his surprise of seeing his mother's double (nine months).[45] Shinn summarized some of these observations at the end of the text:

There is still one important aspect under which these notes should be considered—that is, the development of self-consciousness, which is perfectly evident in both records [of Tilley's two sons]. . . . Once, it is recorded that Laurence watched his thumb move, and was increasingly disposed to investigate his body (9th month, 39th week); once that Winthrop (9th month, 37th week) was perplexed by the duplication of his mother in the glass while taking his own image as a matter of course (since he saw her real self outside the glass, and did not see his own).[46]

But she emphasized the importance of trained interpretation over the simple collection of facts; more important than the exact notation of specific behaviors in front of the mirror was the holistic assessment of a range of behaviors:

> Mainly it is by stopping to think of it that we can trace the process by which the little ones, as they learned to imitate the motions of others, drilled themselves in the difference between the *feel* of the act as seen in another, and as performed by one's self; can realize the wonderful expansion of bodily self-consciousness, as bit by bit, with effort and mastery, the whole body was brought into balance and moved about at will; can appreciate the differentiation between the self and others that grew distinct as communication, by sign and sound, with other human beings, was established.[47]

The mirror test was out of sync with the methodological mores of late nineteenth-century American psychology. It would take another generation and another methodological revolution for the mirror to make a return.

Psychology and Mental Testing

Talbot's and Shinn's networks of researchers sought to turn what might seem an obstacle to scientific research, their responsibilities as mothers, into an asset. But their moment was short-lived because they labored under a scientific regime that, as Robert Nye has shown, increasingly came to gender the researcher as male.[48] Even in the introduction to Talbot's 1882 volume, written by the chairman of her department, William T. Harris, his condescension is all too evident. While Talbot emphasized the importance of a "systematic effort to record the development of infant life" and praised the "hundreds of mothers" who had engaged in the project, "many of who have been trained in our universities and colleges to make investigations with accuracy, and to weigh evidence with candor,"[49] the socially conservative Harris located the value of the project somewhere else altogether: "It does not so much matter what the statistics will show, as it does matter that the mother shall learn to study the growth of her child, and learn what constitutes a stage of progress, and how to discover and remove obstacles to this growth, as well as to afford judicious aid to the child's efforts at mastering the use of his faculties."[50] Harris thus recast a public and scientific project as the expression of the women's maternal duties, with little significance outside the home.

Rather the epistemological values embodied by Shinn and Talbot found their most secure institutionalization in the work of G. Stanley Hall and his "child study movement." Having earned the first psychology doctorate ever awarded in the United States (under William James at Harvard), Hall became a professor at the Johns Hopkins University, where he established a laboratory for experimental psychology, before serving as the founding president of Clark University from 1889 to 1920.[51] Though Hall is most often seen as a Wundtian, he was also deeply influenced by Preyer. He cited Preyer as the main initiator of child study and the predecessor of much of his own research program, and he did so with considerable reverence for Preyer's status as a professor at a German university.[52] In his introduction to the American edition of Preyer's main work, *Die Seele des Kindes*, in 1888, Hall assessed Preyer's text as "the fullest and on the whole the best" of all scientific studies of childhood, "the best example of the inductive method applied to the study of child-psychology."[53] In his own baby diary, which he published ten years after the birth of his second child, in 1891, Hall roughly followed the outline of Preyer's study in the account of his children: development of the senses, development of the will, and a fair amount on language development.[54] Hall was particularly attracted to the way Preyer mobilized his research to answer bigger questions. Building on Preyer's work as well as the work by evolutionary theorists like Herbert Spencer, Charles Darwin, Ernst Haeckel, George Romanes, and Henry Drummond, Hall placed the recapitulation argument at the center of child study in the 1890s, aiming for a synthesis based on evolutionary principles.[55]

Nevertheless, Hall criticized Preyer for his weak methodology, proposing instead ideas drawn from Wilhelm Wundt's experimentalism and a questionnaire methodology, used by German educators from the 1860s onward.[56] In America, as Kurt Danziger has pointed out, Hall's method shifted attention from the study of the individual to populations.[57] Hall's questionnaires left much less room for individual initiative. However suggestive Hall's inquiries might have been, the data he demanded was largely unambiguous. It was not by chance, then, that any inquiries into children's minds that were not as directly accessible were sidelined in Hall's work—much as they were for most of Shinn's network of baby observers. When Hall did conduct studies of this kind, as in his 1895 questionnaire "the early sense of self," we can find only fairly unambiguous indications of self, such as the child's "handling or watching its fingers, toes etc. . . . investigating the dermal boundary of the physical ego," the use of the personal pronoun, and the effects on the child after being praised.[58]

The mirror test is conspicuously absent from the questionnaire, and from Hall's account more broadly.[59]

This absence should not come as a surprise. Hall's epistemological values did not allow the room for interpretation that Preyer left in his account of the *Ichbegriff*. In a brief summary of *Die Seele des Kindes* that Hall gave in the context of a bibliography of child study from 1886 (*Hints at a Selective and Descriptive Bibliography of Education*), Hall sidelined the question of self-consciousness. His entry highlighted Preyer's account of the development of language in part 3 but did not mention the mirror test, which, as we saw, marked the culmination of Preyer's study and the crowning achievement of the child.[60]

While Hall tried to negotiate the uneasy relationship between his Romantic goals and scientific practice, his inheritors in America tended to favor the latter over the former. In particular, Edward Thorndike sought to avoid what he saw as the ambiguity of Hall's questionnaire method (could we really know what lay behind particular answers?) and bring the study of psychology back to the laboratory and the careful, controlled study of animals.[61] For instance in observing how cats, dogs, and chickens escaped from specially constructed boxes, Thorndike came to argue that animal learning proceeded purely by association, rather than by observation, imitation, or reasoning.[62]

The new methodology and statistical methods he introduced helped make child study more acceptable inside and outside psychology. Conceptually and methodologically, child study shifted from the questionnaire to laboratory and quantitative methods. Thorndike's challenge to Hall resonated even at Clark University, the home of the child study movement, and many of Hall's students and collaborators, such as Henry H. Goddard, Edmund B. Huey, Lewis Terman, Fred Kuhlmann, J.E.W. Wallin, and Arnold Gesell, took the new methods on board. Several of the PhDs graduating from Clark produced dissertations on animal learning—on the mental processes of spiders, the mental life of rhesus monkeys, and the psychology of raccoons, to name only a few—of which some used their results to make arguments about genetic psychology.[63]

For many, the best response to Thorndike was a revised form of "mental testing," which aimed to reduce the role played by interpretation and mark individual difference through quantitative means. The mental testers drew on and transformed a longer tradition of "intelligence testing," which could be traced back to the work of Frenchman Alfred Binet in the first decade of the twentieth century. But the test really came into its own only after the 1916 revision by Lewis Terman, another Hall student. In his autobiography, Terman related that, after pursuing a questionnaire study on leadership among children under Hall's supervision, he felt the need to "find a solid footing for research with gifted and defective children."[64] He thus turned to mental tests and did his PhD on the "experimental study of mental tests" under the psychologist Edmund Sanford's supervision, rather than Hall's.[65] He developed what came to be called the Stanford-Binet test. No longer just a means for categorizing feebleminded children, the test came to be widely used by psychologists and in society at large. Putting the entire population on a single scale according to one unified metric of "intelligence," it amounted to a normalization of both the concept and the test itself. Normal intelligence became equal with a certain score on the intelligence scale, adjusting according to age.

Despite the proliferation of the test, it still privileged older subjects over the very young, especially because of its reliance on verbal reasoning. Soon a number of external and internal forces began to push the focus back to the young child, and we will see how this encouraged the reintroduction of the mirror, although with a shifted purpose and meaning. The nursery school movement in the United States, sponsored by middle-class mothers and only later expanded into serving the poor and underprivileged, placed emphasis on the importance of early childhood for the child's later success and ultimately the preservation of American democracy. Hall's former student Arnold Gesell became an outspoken proponent.[66] He made his argument using the language of mental hygiene, the early twentieth-century movement that sought to reduce incidences of mental illness. For Gesell, mental hygiene "concerns itself with the whole developmental span," and although the children's home was still central, both nursery school and kindergarten were necessary to "help to strengthen the home and make up for its deficiencies."[67] More broadly, there was large-scale philanthropic investment in child study during the "decade of the child," the 1920s. At this time, the Laura Spelman Rockefeller Memorial funded Gesell's clinic at Yale, as well as other signature institutions of child study, many of which were concerned with early childhood.[68]

It is in this context that several researchers sought to shift the intelligence test to work with younger children. Rachel Stutsman, who earned her PhD in 1928 under the direction of Helen Thompson Wooley at the Merrill Palmer School in Detroit, noted that the very process of testing required a certain amount of cooperation, self-control, and the ability to understand and follow instructions: "There were no scales of intelligence tests applicable to preschool children except the several revisions of the Binet tests, all of which were scantily standardized and overweighted with tests of a verbal type."[69] And yet,

Stutsman thought that testing would prove to be most useful at an earlier stage of development. An "early diagnosis of general mental defect" could attenuate the "shock" that a diagnosis of "feeble-mindedness" presented to a family, and it allowed participation in early childhood education programs that would mitigate its effects. Moreover, Stutsman expanded the scope to include normative questions that could assess a child's "adoptability" when hereditary factors were unknown.[70]

Mental testers interested in younger children were relatively marginal at the time and for this reason have mostly been passed over in the historical literature. But in their attempt to challenge existing tests, they sought support and engaged in a scholarly conversation across national lines, which linked Charlotte Bühler in Germany and Ruth Griffiths in the UK with a center of gravity in the United States, alongside Stutsman and Gesell, Nancy Bayley, Psyche Cattell, and Mary Shirley. Some of these researchers specialized in cross-sectional tests among a large number of children, while others ran longitudinal studies on just a few. Some conducted the tests in the babies' homes, the rationale being that it was least distracting to test the child in a familiar environment (e.g., Shirley), while others brought the children into their labs.[71] But they all poured great effort into the construction of workable infant tests, which could sidestep issues arising from the child's linguistic limitations. In this way, the mental testing movement produced a rich material culture. A typical testing box included picture books, form boards for doing puzzles, plastic cups and spoons, and various kinds of toys such as a doll, a toy car, and a rattle, as well as a "small, practically unbreakable mirror in a frame" (fig. 2.3).[72]

The Mirror in Mental Developmental Scales: The Loss of the Self

As we have seen, in the child study movement and parallel developments at the turn of the century, the mirror had tended to drop out of the picture. But in the 1920s and 1930s, with the rise of mental developmental testing, the mirror reemerged.[73] Simplicity and accessibility were driving factors. One challenge with testing preschool children was that they would not cooperate if their attention was directed elsewhere. That is why the testing material "should have inherent interest for the child."[74] The child needed to be "tactfully coerc[ed] . . . into doing what is wanted, and for this the attractive apparatus is an ally."[75] The mirror fulfilled this criterion.

FIGURE 2.3. A typical testing apparatus. See the pocket mirror on the left-hand side. The pieces chosen were all "light and easily packed into a small space," so that they could be carried around in the case shown in the photograph. Ruth Griffiths, *The Abilities of Babies: A Study in Mental Measurement* (London: University of London Press, 1954), plate 11, on 119.

But the problem of the mirror remained. For Gesell, "the mirror presents a situation in which the problem of stimulus factors becomes bewilderingly complex." He was particularly unsure what the mirror meant for the child: "How social, how narcissistic, how impersonal, how confused, how illusory, how investigatory, or how naive his [the child's] responses are, we cannot really determine and the risks of interpretation are numerous."[76] Faced with this problem, Gesell decided to sidestep it. Gesell used his apparatus to test for "rattle behavior" (shaking the rattle to produce a sound), "cup and spoon behavior" (mimetic drinking and stirring), or "mirror behavior," defined as a number of different clearly defined actions (smiling at the mirror, approaching the mirror socially, and more).[77] As Gesell's career developed, those behaviors proliferated. In a 1947 paper by Gesell and his colleague Louise B. Ames, "The Infant's Reaction to His Mirror Image," the authors broke down the child's response into sixty "behavior items" listed under six headings: key behavior, emotional attitude, regard behavior, arm-and-hand behavior, body postures and movements, and foot behaviors.[78] This was a considerable expansion from six behavior items for mirror behavior in Gesell's earlier book. But none of these behaviors were thought to indicate self-recognition.[79]

Gesell arrived at these finer categories through the technique of "cinem-analysis," where he recorded on film the mirror behavior of children at the Yale Clinic of Child Development from sixteen to sixty weeks, at monthly inter-vals.[80] Cinemanalysis for Gesell was a "method of observation" that allowed him to better understand the patterns of behavior by analyzing individual frames—similar to Eadweard Muybridge's method of successive photography (fig. 2.4).[81] This could be performed either by slow-motion study (where the film was played at a lower speed), or by what he called "pattern phase analysis," where he selected those phases of the film of greatest importance to him and analyzed the frames in detail.[82] The approach fit with Gesell's broader belief in maturation. For him, "mental growth, like physical growth, is a morphoge-netic process which produces progressive patterns"[83] This growth could be accessed through the "visible forms of behavior."[84]

In all these tests, the trick was to recast the question to allow researchers to give an unambiguous answer. In a chapter of her book The First Year of Life, which was translated into English in 1930 and widely read in the United States, the Viennese child psychologist Charlotte Bühler engaged in a methodological discussion about how to make sense of behavioral data. Interpretation, she was clear, was impossible to avoid. Even a die-hard positivist performed interpreta-tive work, by grouping behaviors together under particular categories.[85] Büh-ler therefore recommended that developmentalists should take Thorndike's cue and place movement in "relationship to its effect or success."[86] Bühler called these units of behavior "acts" or "performances."

Rather than "the listing of each single reflex movement, enumerating and scoring them," the goal was to determine stages of mastery: "A first stage where the child never succeeds; a second stage, where he succeeds sometimes and with effort; and a third stage, where the child is successful always and with ease."[87] What did this mean for mirror behavior? There were various forms of mirror behavior, such as "observing his image in the mirror," or "grasping at the reflection of a cracker in the mirror," which could be marked as "passed" when they were present, and "failed" if there were not. When "the child looks in the mirror and touches and feels of his image in it" or when he "grasps at the mirror in his attempt to obtain the cracker," this earned the child a "+" for that particular test.[88] Bühler believed that Gesell was doing something very similar:

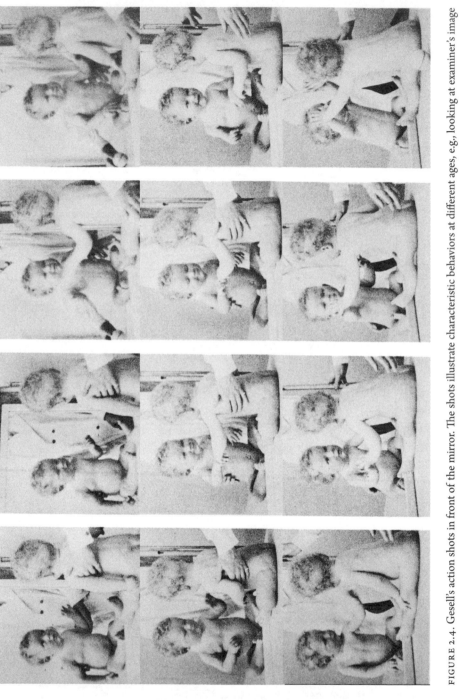

FIGURE 2.4. Gesell's action shots in front of the mirror. The shots illustrate characteristic behaviors at different ages, e.g., looking at examiner's image at forty weeks, patting the mirror at forty-four weeks, and regarding the image of the ball at forty-eight weeks. Arnold Gesell, *An Atlas of Infant Behavior: A Systematic Delineation of the Forms and Early Growth of Human Behavior Patterns*, vol. 1 (New Haven, CT: Yale University Press, 1934), 521.

"When Gesell compares two children, he in fact compares them in their ability to perform certain things." Gesell's "behavior items" for Bühler were really performances.[89]

So too with the Harvard developmental psychologist Psyche Cattell, daughter of the psychologist James Cattell. In 1940 she included two "mirror tests" in her arsenal of infant intelligence tests. One, to be performed at six months was called "Fingers Reflection in Mirror." When the child did what the test prescribed, that is, fingered his reflection in the mirror, credit was given. The other, to be performed at seven months, was called "Pats and Smiles at Reflection in Mirror." It was scored following the same principle. Based on the information whether the child passed these or other tests at specific age levels, his mental age was calculated, following the principle of the Stanford-Binet test.[90]

In an even simpler way, for Nancy Bayley, child psychologist at the Institute of Child Welfare at the University of California at Berkeley, the mirror test, specifically the presence of "mirror image approach" (i.e., approaching the mirror image) was one of 185 developmental test items (number 52, between "turns after spoon" and "picks cube deftly"). Bayley performed each test repeatedly in the sixty-one children she studied, until it was passed, at which point the child was given an aggregate point score.[91] It is in this context that Mary Shirley developed the test with which I began this chapter. At the newly founded Institute of Child Development at the University of Minnesota in the early 1930s, mirror behavior was broken down into different components, each with its own pass-fail test.[92]

The attempts to recast the mirror test into something amenable to the mental testing movement's epistemological ideals served to hollow out its original meaning. First, the tests tended to drift away from the question of self-consciousness. For Gesell and the British child psychologist Ruth Griffiths, mirror behavior was part of the "personal-social category."[93] For both, when certain forms of "personal-social" behavior occurred at certain ages—such as following a moving person with the eyes at three months, stretching the arms to be picked up at six months, or said mirror behavior at ten months—this indicated good social adjustment of the child. Charlotte Bühler classified it under "M" for mental ability.[94] Her tests were designed according to the rationale that they would "measure the single activities which combine to form the main lines of action."[95] Mirror responses were just one of a larger set of behaviors that pointed to four abilities, including: "S" (social reaction), "B" (bodily control), "M" (mental ability), and "O" (response to objects). Bühler considered this information important for producing a developmental profile, which offered a "*distribution* of the child's abilities."[96] Finally, Shirley listed mirror behavior in a chapter on "picture and odor tests," where the baby was presented with pictures, a mirror, and bottles with various smells. Mirror behavior did not seem to indicate anything specific about self-consciousness; Shirley used the same scoring plan for the mirror as for responses to pictures, including reaching for the picture or mirror, turning it to the other side, and involving the examiner into the situation.[97]

In those rare cases where self-consciousness remained an object of study, other changes shifted its meaning. Rachel Stutsman, for example, gave the instruction to present a mirror of about five to seven inches to the child and say to the child, "*Look, who is that?*"[98] If the child responded with its own name, it passed the test. But in relying on the child's own assertion of mirror recognition, Stutsman shifted the test away from the prelinguistic analysis for which it had shown so much promise to earlier mirror researchers, most notably Preyer. Stutsman acknowledged as much: the mirror test was, "at least partially, a test of language development."[99] Another indication that Stutsman did not really take the study of mirror self-consciousness seriously is that, in presenting "sample notes of successful performances," apart from one exception, she did not indicate the child's age when responding.[100] As a corollary, the mirror test stopped performing its earlier role as a demarcator. With the increasing tendency to avoid making claims about what behaviors actually *meant*, it was less easy to see why the mirror should mark human specificity. It is telling that Ruth Griffiths could reestablish language as the intellectual skill "that distinguishes man as man."[101]

Behaviorism

The shifts in the child study movement, from Hall through Thorndike and then to Gesell and others responded to concerns about the baby diary and the need to produce more generalizable data about child development. But they also tracked and were in part driven by the rise of the most important movement in the human sciences in the early twentieth century: behaviorism. Behaviorism, as its 1913 manifesto had it, sought to turn psychology into a "purely objective experimental branch of natural science."[102] This involved three major methodological steps. First, it disqualified introspection, even in the form practiced in the tradition of Wundtian psychology.[103] The behaviorist John B. Watson worried that psychologists used introspection illegitimately to import their own desires and understandings, projecting them on the subjects they studied. Second, and relatedly, it sought to erase the dividing line between humans and animals. Animals could be studied well under experimental conditions, and insights thus gained could be made relevant for humans as well.[104] Finally, this brought the opportunity for psychology to become closer to everyday life, and assume greater relevance for society.

In his work, Watson prescribed a close and hermeneutically cautious description of stimuli and responses. As he explained in his 1919 book, *Psychology from the Standpoint of a Behaviorist*: "*The goal of psychological study is the ascertaining of such data and laws that, given the stimulus, psychology can predict what the response will be; or, on the other hand, given the response, it can specify the nature of the effective stimulus.*"[105] In this way, the psychologist could assure the prediction and control of animal behavior. Watson placed great faith in the possibility of reflex conditioning. He argued that if a particular stimulus appeared regularly during the display of a reflex behavior (the major conditions were frequency and recency), the two would become linked, such that in the future the stimulus could incite the behavior.[106] In a modification of the Russian physiologist Vladimir Bekhterev's conditioned motor reflex, Watson rang a bell while simultaneously applying an electric shock to a human subject's finger, which led to reflex withdrawal. After several repetitions, "in best cases, . . . after fourteen to thirty combined stimulations," he found that the sound of the bell alone produced the same movement.[107] The point here was that any combination of stimulation and response could be produced given the right conditioning; in the study of behavior there was no need to peer into the subject's mind. Watson then applied these principles to humans. In his infamous experiments on Little Albert, which Watson conducted together with his graduate student (and later wife) Rosalie Rayner, they induced in the child a generalized fear reaction to furry entities.[108]

Watson enjoyed a steep professional rise. He became professor of psychology at Hopkins, editor of the *Psychological Review* (the main journal of the field), and director of the psychological laboratory at Hopkins in quick succession. By thirty, he had become one of the most influential figures in psychology. This was in part due to personal circumstances and the forced resignation of his predecessor, Mark Baldwin.[109] But it was also because of the larger social situation. The early twentieth century in America was a time of rapid industrialization, of migration and urbanization, leading to conflict and disorganization. In an era of Progressivism, psychology emerged as a field that promised guidance for the establishment of order and control.

But Watson's career ended as fast as it had begun. After the affair with his student Rosalie Rayner became public, he was forced to leave his post at Hopkins and start a career in advertising. This did not mean the end of behaviorism. Quite the contrary, behaviorism became the mainstream of psychology from the 1920s. And though the figures I have discussed kept their distance from the movement, they also responded to its call for a clearer, less introspective science. Take, for instance, Gesell. On the one hand, he was critical of behaviorism's tenets and especially of its practical applications. He considered behaviorism dangerous for the way it threw "artificial stimuli repeated with artificial frequency" at an infant "in the formative phases of his sensorimotor organization" without taking into account the organism's maturational state.[110] On the other hand, in the same paper, Gesell recognized that the study of the conditioned reflex was a robust scientific tradition using "the most exacting methods," and it is clear that he sought to replicate it in his own work.[111]

Reinstating the Self

Mirror testing from the late nineteenth century to the 1930s had a paradoxical legacy after the Second World War. The mirror test was at once endowed with prodigious meaning, an indicator of something that distinguished humans from animals, and perhaps even pointed to the central mysteries of subjectivity. But at the same time, and in practice, it had become a simple and easy test that elicited a range of well-defined behaviors that were increasingly detached from concepts such as the "ego" or "self-consciousness." In the postwar period, experimenters struggled with this contradictory legacy, trying to construct mirror studies that combined the vaunted goals of Preyer with the methodological rigor of the mental testers. That meant that larger numbers of children needed to be tested under more strictly controlled conditions. Great attention was paid to the kinds of mirrors used as well as their size, setup, and exposure for the child, and testing was performed at more frequent intervals. The French psychologist Geneviève Boulanger-Balleyguier, for example, tested the mirror response of thirty infants, under various conditions, every month, over a period of half a year.[112] Gone was the time when a homely atmosphere was considered important; now, the goal was to reduce distractions. Studies emphasized the "emptiness" of their testing rooms, which were often located in a laboratory: "The room, 4.15 m × 5.5 m, was painted white and beige and was empty except for a chair and the video equipment which was placed on a large desk."[113] Experimenters paid careful attention to the setup. The distance of the mirror from the child, the kind of mirror used (flat, blurred flat, distorted, blurred distorted, etc.) had to be carefully recorded and testing conditions held stable.[114]

Some researchers drew on new technologies to enhance the experiments' rigor. Video was seen to have several advantages, the most important being its ability to shift and test variations on the mirror. The video could be played back to the child at the same time as it was recorded or with a short delay. This foregrounded for researchers the question of synchrony and its significance for self-recognition: Did the child recognize its image because that image moved in concert with its own body? In a study conducted at Simon Fraser University in the late 1970s, Ann Bigelow exposed children to mirrored images under several different conditions: simultaneity of movement between self and image, in mirror and video conditions; and nonsimultaneity of movement between self and image, in video discordant and photograph conditions.[115] Self-recognition was noted either when a child turned around after seeing a clown face in the mirror, or when it identified the image verbally, or both. Bigelow found that children recognized themselves earlier in the simultaneous conditions than in the nonsimultaneous conditions.[116] But the findings were far from uncontested. Other authors doubted the importance of the role of synchrony. The physicians and developmental psychologists Hanuš and Mechthild Papoušek, as well as the child psychologists Beulah Amsterdam and Lawrence Greenberg, found that the visual cue of eye-to-eye contact between the infant and its mirror image was more important than synchrony in the process of developing mirror self-recognition. Children at the age of five months showed preference (as measured by visual attention to the screen) of a televised movie of themselves over their TV mirror image.[117]

The question of synchrony, and thus the process by which the child came to recognize the mirror image, refocused attention on the sequence of mirror behaviors that had been so important to Preyer, and many embraced a stadial model. As the psychologists Bennett Bertenthal and Kurt Fischer wrote, "Investigators need no longer argue about which behavior is true self-recognition. Instead, they can simply specify which stage or stages of self-recognition behavior they are examining."[118] Others worked hard to describe precisely what the stages of self-recognition were. Testing a group of infants between the ages of four and twelve months, J. C. Dixon, a developmental psychologist at the University of Florida, distinguished four stages of recognition in 1957.[119] In the first stage ("Mother"), around four months, the infant might look briefly at its own reflection without showing any sustained interest. In contrast, the mirror image of the child's mother, was regarded with great attention. In the second stage ("Playmate"), from about four to six months, the child's behavior toward its mirror image was the same as that shown toward another baby. The third stage, between six and seven months, misleadingly[120] called the "Who dat who do dat when I do dat?" stage, the child showed repetitive action in front of the mirror, discovering and playing with the fact that the child in the mirror behaved exactly the way the child did. And finally, in the fourth stage (named "Coy"), we see a return to the emotional responses that nineteenth-century observers had noted in their children: between twelve and eighteen months, children turned their heads away from the mirror image. Some would cry, while others would smile coyly and then look away.

Here as well, controversy reigned. Amsterdam was skeptical about Dixon's "Who dat who do dat when I do dat" phase (a phase the core of which—behavior synchronicity—was the focus of the video experimenters already discussed). Rather, she believed that at that stage of development the child still saw the mirror image as another child. This, she pointed out, was evident from the data Dixon himself presented, where the child "treats his image as a 'playmate.'"[121] Like Dixon, however, Amsterdam thought that mirror self-recognition did not occur out of the blue. Rather, she emphasized that the shift from seeing the image as another child to seeing it as oneself was *mediated* by the phase of emotional withdrawal: "In support of the hypothesis that the child has some awareness that he is looking at his own image is the evidence provided by the embarrassed self-consciousness and avoidance behaviors associated with locating the red spot and the verbal self-recognition response. Every subject who showed recognition behavior also manifested either avoidance or self-consciousness [in the sense of embarrassment], or all three. Recognition inevitably appears as one element in a complex pattern of behavior."[122]

The "red spot" in the last example refers to what came to be called the "mark test," which had been developed by Amsterdam as part of her 1968 PhD dissertation at the University of North Carolina at Chapel Hill in response to her intense dissatisfaction with the lack of agreement about mirror behavior and especially about the "type of behavior that indicates clear self-recognition in children."[123] Amsterdam therefore set herself the task of devising "a non-verbal technique which could be used in a standard manner to study self-identity in children under two years of age."[124] In the test, a spot of rouge was placed on the child's nose without the child noticing it. The child was then placed in front of a mirror. If the child, after examining its mirror image, either reached for the rouge on its own face, or used the mirror to examine its nose closely, this suggested that it had understood that the mirror image was a reflection of its own body, and thus that it had recognized itself.[125]

As we will see in chapter 4, the mark test had been developed independently by the comparative psychologist Gordon G. Gallup at almost exactly the same time to demonstrate self-awareness in chimpanzees. And though the mark test brought its own set of difficulties, it was regarded by many to be an ingenious solution to the problem that had animated so much of the research over the past eighty-five years: the ambiguity of mirror behavior.

But before we leave the moment marked by the tension between the aspirations of the mirror test and behaviorist or behaviorist-adjacent methods, it is worth moving away from the rather broad tradition we have examined in this chapter and zooming in to examine a single figure who brought out most clearly its contradictions: William Grey Walter. His work is interesting to us for another reason. While the other scientists we have studied deployed the mirror to examine the moment in childhood development when distinctively human traits first emerged, or to draw the line between humans and nonhuman animals, Walter used it in his work at the boundary separating humans from robots.

3

The Dancing Robot

GREY WALTER'S CYBERNETIC MIRROR

IN A FAMILY HOUSE near the English city of Bristol, a metallic object—about the size of small dog, with a rotating periscope-like protuberance peeking out of a shell at one end—wanders in loops around the living room. In its sweeping itinerary, it is faithful to its name: *Machina speculatrix*, the exploring machine. At the end of one cycle, it finds itself abruptly in front of a large mirror, and its movement changes decisively. It no longer seems to be exploring; rather it moves in a zigzag pattern, scanning across the mirror before it finally enters its "kennel" (fig. 3.1). As its creator, the British cybernetician William Grey Walter, noted, this was a "mirror dance," an "absolutely characteristic mode of behavior, which is seen always and only when the creature is responding to its own reflection."[1] It "lingers before a mirror, flickering, twittering and jigging like a clumsy Narcissus."[2]

Walter then indulged in a thought experiment:

> Put yourself in the position of a biologist exploring an unknown island. He comes across a hard-shelled creature; when he puts a mirror in front of it, it behaves in a specific way. He would write a letter to *Nature* and say he had evidence of recognition of 'self' on the part of his mollusk, crustacean, or whatever it was.[3]

The scene and the imagined reaction to it are not too dissimilar from the many others we have encountered in this book: The peculiar behavior of a creature in front of the mirror is interpreted as evidence of some crucial cognitive capacity. But Walter was less participating in the tradition than parodying it.

Walter was not mocking any particular version of the test. He cited no other mirror tester, and, after all, at the time he was writing, no one had claimed that

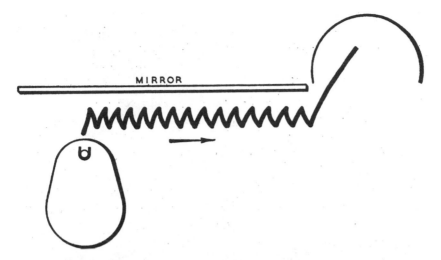

FIGURE 3.1. The tortoise's mirror dance. W. Grey Walter, "Presentation: Dr. Grey Walter," in *Discussions on Child Development*, ed. J. M. Tanner and Bärbel Inhelder, vol. 2 (1956; New York: International Universities Press, 1971), 21–74, on 35.

a nonhuman animal could recognize its reflection. Rather his joke arose from the collision between a broader belief that the mirror test could demonstrate self-recognition and an understanding of its epistemic weaknesses that was available to anyone who tried it. Grey Walter was playing on the mismatch between these two, something that was particularly apparent in his case. Whereas earlier researchers used the test as one of the only possible ways of gaining access to the mind of an animal, Grey Walter knew exactly what was going on in the head of his "tortoise." After all, he had designed and built it. However suggestive the behavior, Grey Walter knew that the machine could not have recognized itself, because he had not provided it with the wherewithal to produce and hold anything like a concept, let alone a self-concept.[4] In this way, Grey Walter's cybernetics performed the same role played for Wilhelm Preyer by neurology, though to opposite effect. It allowed him to peer into the black box of the mind, not to be better able to read the subject's behavior, but to discount it.

Walter's tortoises have attracted significant attention, from the public and historians alike. They were the subject of a BBC documentary in 1950, "Bristol's Robot Tortoises Have Minds of Their Own," which takes the viewer into the Walters' home near Bristol, where the tortoises had been constructed and where they were let loose to show their tricks with a smiling Mrs. Walter

FIGURE 3.2. The two tortoises in the exhibition catalogue for the Festival of Britain. The caption reads: "Mechanical 'animals' can be made to steer themselves towards the light." Note that one of the tortoises was presented as "looking" into a mirror. 1951 Exhibition of Science, South Kensington, Festival of Britain, Guide-Catalogue. Dr. W. Grey Walter. Advice and supervision of Model of Electrical Tortoises. With kind permission from the National Archives, ref. WORK25/230/C1/A1/3.

standing in the background.[5] The following year, the *Machina speculatrix*, also called Elsie (short for **e**lectromechanical robot, **l**ight **s**ensitive with **i**nternal and **e**xternal stability), and her brother Elmer (**e**lectromechanical robot, **l**ight-sensitive) were on display at the Festival of Britain, a national exhibition in the summer of 1951 (fig. 3.2 and plate 3).[6] In the exhibition, the tortoises were shown frolicking in a pen.[7] They responded to lights installed at floor level, which could be switched on or off by the visitors.

The response was enthusiastic. As *Punch* magazine noted, the robots, which looked like "a couple of electric kettles crawling about the floor, the spout being replaced by a revolving periscope," were one of "the two best toys in the exhibition."[8] Public interest, having peaked in the mid-twentieth century, is still present, even outside Britain, and a descendent of Elsie and Elmer is part of the collection at the Smithsonian.[9] Certainly Walter's showmanship and maverick status, as the historian of science Rhodri Hayward has observed, helped the popularization of his devices.[10] But the excitement was primarily due to the impression that Walter had created mechanical life, breaking down the distinction between the living and the inorganic.

Scholars today share this fascination. The computer scientist Owen Holland has painstakingly reconstructed the details of the tortoises' workings. Interested in intelligent autonomous systems himself, he built replicas of the machines in order to appreciate their rich behavior enabled by a small number of "brain cells."[11] Historians of science have placed more emphasis on the complex connections between the robots and larger concerns in Walter's career, as well as the broader medical and cybernetic context.[12] In both cases, these scholars seek to gain insight into the robots' strange lifelike behavior. As Andrew Pickering has argued in *The Cybernetic Brain*, Walter provided "sketches of another future," an entirely new way of understanding our relationship to the world: "The assertion that the tortoise, manifestly a machine, had a 'brain,' and that the functioning of its machine brain somehow shed light on the functioning of the human brain, challenged the modern distinction between the human and the nonhuman, between people and animals, machines and things. This is the most obvious sense in which Walter's cybernetics, like cybernetics more broadly, staged a nonmodern ontology."[13]

Pickering's list of surpassed oppositions is intended to demonstrate Walter's radicalism; here was a man who refused to be confined by the constitutive dualities of the modern age. But in his enthusiasm for Walter's work, Pickering elides conceptual pairs that for Walter were importantly distinct. As I will show, Walter's project of breaking down the distinction between the human and the machine was part and parcel of a larger project aimed at reinforcing the human-animal divide. These two projects collided in the mirror test, where the borders between the human, the animal, and the robotic were brought into play.

Cybernetics in Britain: Walter, the Ratio Club, and the Fascination with the Brain

The person usually credited as the founder of cybernetics is MIT mathematician Norbert Wiener. Wiener coined the term—which means literally something like the "science of steermanship"—and published his book *Cybernetics; or, Control and Communication in the Animal and the Machine* in 1948. The book covers a broad swath of topics, with chapters on time, communication, psychopathology, and "Information, Language and Society."[14] This corresponded to the trend set by the famous Macy cybernetics conferences held between 1946 and 1953 in New York, which brought together a dizzyingly interdisciplinary array of scientists. Despite the wide variety of the fields it engaged, cybernetics boiled down to a single guiding idea: regulation through feedback systems. As Wiener pointed out in the introduction to his book, feedback was a way of predicting the future, with very useful consequences: "When we desire a motion to follow a given pattern, the difference between this pattern and the actually performed motion is used as a new input to cause the part regulated to move in such a way as to bring its motion closer to that given by the pattern." For example, when coupled with the tiller through a feedback mechanism, a steering wheel in a ship was able to chart a stable course, which was "relatively independent of the load."[15]

Walter too traced the origin of cybernetics back to America and Wiener, but he nonetheless carved out space for an independent and, in his view, superior British tradition.[16] In a letter to his fellow British cybernetician John Bates, Walter discussed an encounter with Warren McCulloch, a pioneer of the American movement. Walter shared his impression that McCulloch's "thinking had reached some sort of plateau," if, he added generously, at "quite a high level." Bates responded that he found "all Americans less clever than they appear to think themselves."[17] Moreover, as Bates wrote at another time, the British had come up with "Wiener's ideas before Wiener's book appeared."[18]

The British cybernetics movement was complex and multifaceted, but its core was an unofficial discussion society named the Ratio Club, which was active between 1949 and 1955.[19] Though Britain had been exhausted by the war, and was reluctantly engaged in an ongoing process of decolonization, this was a time of burgeoning optimism and hope in science, which provides us with a lens through which we can understand Walter's work. In particular, it draws attention to two defining features of the British movement. First, British cybernetics depended on a deep relationship with the contemporaneous brain sciences, and second, it celebrated a makeshift approach, often presented as a legacy from the Second World War.[20]

The emphasis on the brain in the British tradition is clear from the list of founding members of the Ratio Club, which included, in addition to Walter and Bates (who was a neurologist at the National Hospital for Nervous Diseases), a sensory physiologist (Horace Barlow), a psychologist (W. E. Hick), and a statistical neurohistologist (D. Scholl), among others.[21] The very location of the Ratio Club pointed to its proximity to the neuro disciplines. With Bates's help, the group secured a basement room "below nurses' accommodation"[22] at the National Hospital for Nervous Diseases in the Bloomsbury district of London. Neurological expertise was then brought to bear on the club's cybernetic work. The members constructed machines by modeling them on the nervous system, all part of the larger cybernetic project of the "mechanization of mind."[23] The close relationship the Ratio Club constructed between brains and machines is visible in the list of possible discussion topics, drafted by psychiatrist William Ross Ashby on February 18, 1950: we find themes ranging from "'noise' in machines and brain" through "pattern-recognition in machines and brains" and "free-will" to "the diagnosis and treatment of insanity in machines."[24] These were brought together in a laddish and boozy atmosphere. At the end of the document, Ashby proposed a final discussion topic: "If all else fails: The effect of alcohol on control and communication," a discussion that would inevitably involve "practical work."[25]

If the emphasis on brain science can be explained by the Ratio Club members' professional concerns, their make-do attitude was due, at least in their self-presentations, to the experience of the war. In their wartime service, many had acquired additional skills in the physical sciences or engineering; for example, Thomas Gold and John Pringle, zoologists at Cambridge University, were "ex radar" workers.[26] This wartime experience was also closely related to what many saw as the peculiarly "British" nature of their work. Just as they had been compelled to develop clever and easy solutions to technical problems under the threat of Nazi Germany, so too they praised ingenuity in their peacetime work.

The favored contrast of British ingenuity in the face of the American brute power is exemplified by an anecdote Walter liked to tell about his relationship to Wiener. In the early postwar period, neurophysiologist Mary Brazier, who worked with Robert Schwab at the Massachusetts General Hospital, had summoned Walter to Boston to share his insights about EEG (electroencephalography) frequency analysis.[27] Walter thus found himself only a short distance from Wiener, who was working at MIT on similar problems. In order to break the complex brain wave down into its constitutive parts, Wiener performed a rigorous Fourier analysis. The process was mathematically taxing and required considerable time. Walter in contrast had constructed a machine consisting of a set of tuned reeds that vibrated at particular frequencies. When connected to an EEG current, each reed hummed if the signal contained frequencies matching its own. In this way the machine provided a rough but instantaneous analysis of the wave.[28] As Walter recounted in his 1969 "Neurocybernetics," a contribution to the collected volume *Survey of Cybernetics: A Tribute to Dr. Norbert Wiener*, "I was quite put out to hear Wiener holding forth about the theory and principles of frequency analysis applied to brain waves as if this were a novel and difficult concept, when my machine was ticking away almost next door, reeling off brain wave spectra automatically, every ten seconds, hour after hour."[29]

Walter and the EEG

Like that of other members of the Ratio Club, Walter's cybernetics should be seen as an outgrowth of his neurological research. Despite their public profile, the tortoises were more a pet project than Walter's main professional concern. He constructed them in his free time at his home on the outskirts of Bristol and studied their behavior in his living room, walking after them in his slippers. He published little on the robots—the best-known accounts are one chapter in his book *The Living Brain* and a pair of articles in *Scientific American*.[30] But Walter did work extensively on neurology, and more particularly EEG—at the Burden Neurological Institute near Bristol, where he was director of physiological research—which was the subject of the great majority of the 174 scientific papers Walter published during his lifetime.[31] Walter's robots can be understood only within this broader context of Walter's professional life.

That Walter would dedicate his career to EEG came as something of a surprise. As a student of the physiologist Edgar Adrian at Cambridge—passing the natural science tripos in 1931, and then staying on until 1934 working with Bryan Matthews in Adrian's department—Walter had rather focused his attention on lower functions. His master's thesis with Adrian was "Conduction in Nerve and Muscle," in line with Adrian's broader research program on the electricity of nerves (Adrian discovered the "all or nothing" function of nervous action potentials, for which he was awarded the Nobel Prize in Physiology or Medicine in 1932). For Adrian, the human brain was out of bounds for a self-respecting physiologist. If Adrian did stray beyond the spinal cord, it was almost exclusively in the study of animals, not humans.

Walter's path to the brain was opened up by an innovation from Germany. Starting in 1929, Hans Berger at Jena published a series of papers on what he called *Elektrenkephalogramm* (electroencephalogram), a method of picking up the electrical activity of the brain.[32] Every time a neuron fires it produces a tiny electrical charge. The concerted firing of multiple neurons in one region could produce a signal large enough to be measured by electrodes attached to a nearby area of the scalp. If these electrodes were connected to an amplifier, which in turn was hooked up with a recording device that would register the electrical changes over time on a piece of paper, the "brain waves" could be traced and studied. As one of his first results, Berger showed that when a subject was at rest with her eyes closed, the brain produced a wave of about ten cycles a second, what he called "alpha waves." The alpha waves would disappear when the subject opened her eyes.

For Walter, Berger's results opened up entirely new vistas of research. The EEG showed that the "deus ex machina" of electricity, which had been identified as the working principle of the nerves by scientists like Adrian, had now been transferred to the cortex. He speculated that "the physiological background of our perception and thinking may be neither peace nor chaos but a deep all-embracing rhythm of which we are unconscious, perhaps because we are so used to it."[33] As Walter saw it, the EEG was the royal road to the brain.

It was perhaps a similar enthusiasm that led Adrian and his coworker Matthews to take interest in Berger's work; they could use it to put their recent findings about animals and electricity to the test.[34] In their experiment, Adrian and Matthews refined Berger's technique. German researchers had used multiple recording electrodes, but because these had been routed through a single amplifying system (e.g., in the work of Jan-Friedrich Tönnies working in Berlin), it produced a single output, one wave for the entire brain.[35] In contrast Adrian and Matthews used three amplifier systems, each attached to separate sets of electrodes, which introduced a new possibility into EEG research: localization.[36] Because the amplitude of a signal was dependent in part on the distance between the wave's focus and the electrode, the comparison of multiple recordings could indicate where the waves originated.[37] On the basis of the experiment, Adrian and Matthews were able to locate the focus of the alpha waves (they termed it the "Berger rhythm") in the occipital lobe.[38] Despite this success, Adrian and Matthew did not take EEG research further. As the historian Cornelius Borck has noted, the EEG lay outside Adrian and Matthews's experimental system, and so they did not feel compelled to follow up on its possibilities. They merely noted that the rhythm was "disappointingly constant," expressing a time relation that was characteristic for neurons in general.[39] Walter then picked up where his teachers had left off.

Walter's opportunity to do this arose, ironically, from his failure to win a college fellowship for doctoral study at Cambridge. In 1935 he began postgraduate work with Frederick Golla, the director of the Central Pathological Laboratory at the Mental Hospital in London. Four years later, they moved together to the Burden, where Golla took the position as director (fig. 3.3). Unlike Adrian, Golla encouraged Walter's interest in the EEG, and in the period before the outbreak of war in 1939, Walter used it to study brain tumors. As Walter discovered,

FIGURE 3.3. Burden Neurological Institute (BNI) in 1939, the year it was founded. The institution housed both clinical and research departments. Archives of the Burden Neurological Institute, Burden A/6/19 (Papers held by Dr. Ray Cooper).

tumors produced an electrical signal with a lower frequency than the alpha waves, which he termed "δ [delta] waves."[40] He explained the name in *The Living Brain*: "delta rhythm" seemed fitting "because of its association with disease, degeneration and death," as well as "defence."[41] As the final association suggests, the discovery had immediate clinical implications. Using Adrian and Matthew's technique, Walter was able to use the EEG to locate a tumor to "within about four centimetres," thus sparing the patient a dangerous exploratory surgery.[42]

War Technologies

Walter was ambivalent about the clinical orientation of his research. Despite its obvious value, it was focused on pathology and not normal function and thus did not have for Walter the same implications for understanding how the brain worked. Alpha waves, the product of a normal healthy brain, were for him a much more promising subject of research.[43] The opportunity to think about the latter was a curious side effect of the global conflict that broke out in 1939.

The impact of the war on Walter's understanding of the nervous system can be accounted for in two ways. First it gave birth to a project of modeling brain functions with machines. During the war, Walter remained at the Burden, helping the neurologists and neurosurgeons with their work and also administering ECT to those diagnosed with "'battle fatigue,'" with Walter's own scare quotes.[44] He credited his new insights into the nervous system to his exchanges with the Cambridge psychologist Kenneth Craik, who was in the thick of the technological struggle against the Nazis.[15] Craik, appointed to various war committees including the Military Personnel Research Committee and the Target Tracking Panel (which he chaired), developed a range of approaches to understanding how humans operated in combat settings. For instance, he created machines for investigating fatigue and visual adaptation. He also worked on modeling human operators in control systems to optimize their functioning; in this, he followed the basic assumption that "*the human operator behaves basically as an intermittent correction servo.*"[46] Craik's mechanical simulacra of airmen and soldiers can be seen as direct antecedents to Walter's tortoises.

Second, even war technology not directly mobilized to understand the human brain could be useful. For Walter, the brain was a highly complex organ, but one that for evolutionary reasons operated according to Occam's razor, seeking the simplest and most efficient means of achieving any result. Such ingenuity was also demanded by the extraordinary conditions of war, and thus it is not surprising that clever solutions to difficult problems produced in the war effort might find parallels in the working of the nervous system.

One such technology was scanning.[47] Walter remarked on the difficulty of transmitting highly complex images from one part of the brain to another. As he posed the problem, the metaphorical director in the brain, who was too busy to "go to the projection room" himself, needed a means to work out what was happening there.[48] How was it possible that visual stimuli could be worked on by parts of the brain beyond the visual projection areas? Or, in his words: "How is the spatial image which is received in the projection areas transferred to the other areas for cognition?"[49]

Walter put aside the idea that the image could be transmitted point for point; that "would require more space than the largest animal could carry" and would also make it difficult to compare the visual data with that from other senses.[50] Rather, Walter drew inspiration from scanning machines, most notably radar.[51] In a radar system the receiver rotates, sweeping across 360 degrees, such that the location of an object in space was translated into a function of time. This approach transformed two-dimensional data (coordinates) into a

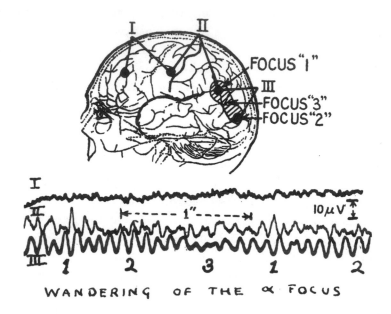

FIGURE 3.4. Wandering of the alpha focus from under the parieto-occipital electrode (FOCUS "1") to under the occipital electrode (FOCUS "2") and from there to a point between the two (FOCUS "3"). W. Grey Walter, "The Electro-encephalogram in Cases of Cerebral Tumour," *Journal of the Royal Society of Medicine* 30, no. 5 (1937): 579–98, on 586. Reprinted by permission of SAGE Publications.

single signal, which could be transmitted economically. Walter hypothesized that the alpha waves might derive from a similar process. As a by-product of his experiments on delta waves, he had noted that the focus of alpha waves wandered (fig. 3.4), from somewhere under the parieto-occipital electrode (FOCUS "1") to under the occipital electrode (FOCUS "2") and from there to a point between the two (FOCUS "3"). That is, the alpha waves seemed to be "scanning" the visual projection area like a radar.

The idea that the alpha wave was a form of scanning was confirmed for Walter by an experiment he performed with a flickering light.[52] Walter and his colleagues noted the "peculiar effect" of "a vivid illusion of moving patterns whenever one closed one's eyes and allowed the flicker to shine through the eyelids."[53] When the eye received stimulation from a flickering light at about the same frequency as the alpha waves, it produced a number of visual hallucinations including the "appearance of a mosaic or chessboard pattern, sometimes with a whirlpool effect superimposed."[54] The illusions could be explained by a scanning system. With a low flicker rate, each flash was seen as a single event: the alpha wave could complete its scan before the next flicker started, and "the EEG shows a single evoked potential change in the occipital cortex."[55] But the scanning always took a certain time. If the flicker flashed more quickly, such that the light would be both on and off during a single scan of the image, the radar would read it as complex and changing. As Walter presented it in a talk given to the Ratio Club: "Scanning systems . . . require time and are upset by intermittent signals—they achieve good resolution with great economy of channels but generate illusory patterns when the signals are discontinuous functions of time."[56]

The best mechanical model for the alpha wave, however, was not the radar but rather the "line-scanning device," "which employ[s] a system of error-operated positive feed-back."[57] The technology had been developed to control self-guided missiles and thus came with all their promise and threat. Much like a self-guided missile, a scanning device when diverted from its path, will find its way back on its own. As Walter wrote: "If you have a scanning device, a rotating photo-cell, then once it gets on to a line it will tend to pick up that line again within reasonable time. The same unfortunately is true of self-guided missiles If a missile is aimed at London and interrupted by some counter-force, it is easy to make it take up the line again, after the perturbing force has been circumvented. This is characteristic of quite a simple system without storage."[58]

The line-scanning device functioned in the following way. A cathode ray produced a spot or light on a screen. Facing the screen, a light-sensitive photocell was connected both to an inscription device and, via an amplifier, to an electromagnet. On exposure to the light dot, the sensor activated the magnet, deflecting the ray of electrons and causing the dot to rise until it reached the edge of the screen. If an opaque mark were placed on the screen, however, the dot would remain at its edge, because if the dot rose beyond it, it would be obscured, turning the electromagnet off and thus causing the dot to sink until it reappeared. When this apparatus was combined with a lateral scanning function, the light dot would oscillate until it hit the edge of the opaque shape on the screen, at which point the dot would sketch out its outline (fig. 3.5, *top*).

In this way the line-scanning device reproduced the "peculiar properties" Berger had noted in the alpha wave, which is why he placed an alpha wave EEG underneath the line-scanning signal for comparison (fig. 3.5, *bottom*): (1) the alpha wave "dies away when the visual mechanisms are most active" (when the basic scanning picks up the opaque shape); (2) it "occupies a limited area in

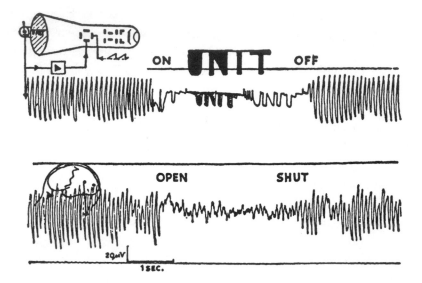

FIGURE 3.5. *Top*: Line-scanning device. The silhouette of the word "UNIT" was attached to the screen, which the light dot (and hence the output—the wave below) sketched out. *Bottom*: EEG of alpha waves, with eyes closed and then open. From W. Grey Walter, *The Living Brain* (New York: Norton, 1953), 110. Copyright 1953, © 1963 and renewed © 1981, 1991 by W. Grey Walter. Used by permission of W. W. Norton and Company, Inc.

the projection and association visual fields"; and (3) "it is usually complex and should really be spoken of in the plural." Walter continued: "Its constancy may have disappointed Adrian, and indeed its frequency cannot vary by more than 10 or 20 per cent, but within these limits it is quite sensitive not only to psychological but to biochemical and pathological disturbance."[59]

"Tortoise Because He Taught Us"

This was Walter's motivation for creating his tortoises, building them out of discarded war electronics and applying the "British" emphasis on technological ingenuity to their construction. At the Festival of Britain in 1951, Walter's tortoises were listed under the How We Know section, which focused on such neurobiological topics as sensory perception and the anatomy of the brain and nervous system (fig. 3.6).[60] The tortoises were intended to illustrate the process of automatic or reflex action: "The senses send their findings rather like electric signals along the nerves. Such a signal may set off an automatic or reflex action; this is how a shadow across the eye makes us blink, or an insect

FIGURE 3.6. The tortoise pen (without the tortoises) at the Festival of Britain. Note the model of reflex action on the right. Dr. W. Grey Walter. Advice and supervision of Model of Electrical Tortoises. Reproduced with kind permission from the National Archives (UK), ref. WORK 25/214 (4922).

moves towards the light. A mechanical 'animal' can be constructed to steer itself towards the light in this automatic way."[61]

The *Machina speculatrix* was Walter's simplest robot. It was constructed out of the interconnection of two receptors and two effectors, in this way modeling a simple reflex (that is, sensory-motor) model. One receptor, a photoelectric cell, also known as the machine's "eye," was coupled with one of the effectors, the single driving wheel at the front of the robot, so that both "always face[d] in the same direction."[62] The other effector was the steering motor, which produced a continuous rotating movement of the photoelectric cell (and consequently of the driving wheel) so that it "scans steadily."[63] The second receptor, connected to the outer shell, recognized when the tortoise ran into an obstacle or met with a gradient. The relationships between these four elements were governed by certain rules: (1) The scanning (and steering) rotation would come to a halt whenever the photocell was exposed to moderate light. (2) It would pick up again at half speed when the light intensity reached a specific level. (3) The driving motor would run at half speed when the photocell recorded a low level of light but at full speed when it detected moderate and intense light.

The behavior resulting from these simple rules was, as Walter had intended, highly complex. A basic pattern can be identified: phototropism (fig. 3.7). If the tortoise were placed in a room with a single light in the middle, the light sensor would rotate with the motor running at half speed, leading the tortoise on a cycloid path. At a certain point the sensor would point toward the light in the middle of the room, activating it. The scanning mechanism thus would stop, and the driving motor would rise to full speed, careening the tortoise toward the light. Once it was close enough to the light, however, the light sensor would reach capacity and the scanning would turn on again, leading the tortoise to look away from the light, and thus to slow down.

As Walter remarked, he named the machines "tortoises" in reference to *Alice in Wonderland*. Walter quoted a dialogue between Alice and the Mock Turtle:

> When we were little . . . we went to school in the sea. The master was an old Turtle—we used to call him "Tortoise." "Why did you call him tortoise if he wasn't one?" Alice asked. "We called him tortoise because he taught us" said the Mock Turtle angrily: "really you are very dull!"[64]

The tortoises, for Walter, could teach us about the building blocks of the nervous system. If the alpha wave was like a scanning device, then a scanning device, if correctly constructed, might be able to mimic certain functions of the brain. It is true that the tortoises were not animals. As machines, Walter confronted the presumption that they were fundamentally, even ontologically different from the creatures they were supposed to mimic; he worried that their behavior might be regarded simply as a "trick."[65] Nevertheless, he was adamant that this wasn't the case. As Walter wrote, "The observation of these models has the same degree of validity as the study of animals, if you are relating your studies to human problems. In studying animals one is studying what one supposes to be a model of human behavior, and one is perfectly at liberty to anthropomorphize—'the animal does so and so, and this is what I feel when I do the same thing'—if it makes thinking clearer and more vivid and makes the hypotheses more precise and conclusive."[66]

FIGURE 3.7. Phototropism. Elsie moved toward the candle but also demonstrated "discernment" by avoiding the obstacle on the way. Owen Holland, "The First Biologically Inspired Robots," *Robotica* 21 (2003): 351–63, on 361. Reproduced with permission from Cambridge University Press.

This argumentation might be called an "inverse behaviorism." As we have seen, behaviorism required a single-minded focus on animal behavior as a guard against our inclination to identify with their actions and thereby attribute to those animals the internal states we experience as humans. Walter too argued that we should put aside our intuitions, and focus on behavior. But because he was dealing with robots, the situation differed in two important ways from the one that had produced Watson's method. First, and most importantly, the robot did not present a black box to the researcher. Walter was able both imaginatively and literally to enter into the internal processes of the robot because he had built it. Second, the intuitive assumptions Walter wanted to stymie weren't those that constructed an unjustified identification between animals and humans. Rather they were those that refused any identification between humans and robots because of prejudices about the nature of thought. Though Walter used the term *anthropomorphization*, what he was actually arguing for was identification in the other direction, projecting cybernetic mechanisms onto organic processes. If a robot and an animal presented the same behavior, then Walter thought we could assume that the electronic structure of the robot mimicked the neurological structure of the human. Robots could tell us something about how biological brains worked.[67]

This argument could be directly applied to the alpha wave. The scanning of the light sensor was the core of the robot's workings. Much like the alpha waves running in the background as a default activity when the human subject was not engaged in any particular task (visual or other), in its rest state the robot would steadily scan its environment and, through the coupling with the driving wheel, spiral around the room. Like the alpha scanning, the machine's scanning would stop under a certain condition: a particular visual stimulus (moderate light). And like the alpha scanning, the tortoise's scanning would be taken up again when the stimulus went away. The tortoise also mirrored the supposed function of the alpha waves, because it transformed visual stimuli (in this case the light) into a signal, which could be sent to another part of the organism (in this case the driving wheel). As Walter wrote: "This process of scanning and its synchronization with the steering device may be analogous to the mechanism whereby the electrical pulse of the brain known as the alpha rhythm sweeps over the visual brain areas and at the same time releases or blocks impulses destined for the muscles of the body."[68]

Walter gained confidence that he had successfully mimicked the alpha wave through the *Machina speculatrix*'s behavior. The reconstructed alpha wave at the heart of the tortoise's construction had produced a locally random but globally directed behavior: seeking out and flitting around the light. Thanks to such behavior, Walter came to understand his tortoises to be dramatically different from that other machine invested with significant hope by cyberneticians: the computer. Computers were, Walter admitted, much more complex than the tortoise, and this allowed them to be as good as, even superior, to human beings "in their great speed of action and in their ability to perform many interdependent computations at the same time."[69] Like the calculating machines designed by the

English mathematician and inventor Charles Babbage, these machines never "lost count."[70] In comparison, the tortoises were "less ambitious" being several orders of magnitude simpler than human brains; while the latter had some "10,000 million" brain cells, the tortoises only had two functional elements.[71]

But for Walter, for all their size and power, computers could provide very little insight into the brain; their behavior was predictable, even "predetermined."[72] As he wrote, "Their resemblance to living creatures is limited to certain details of their design. Above all they are in no sense free as most animals are free; rather they are parasites, depending upon their human hosts for nourishment and stimulation."[73] The machines that "Wiener foresees" had a similar problem: They "may entirely supplant human labour in the factory, but it will be of little interest to the physiologist. It will no longer be part of a mirror for the brain."[74]

The tortoises were different. Through them, Walter thought, we could "discover what degree of complexity of behavior and independence could be achieved with the smallest number of elements connected in a system providing the greatest number of possible interconnections."[75] They had demonstrated a qualitative step beyond predetermined behavior, one that seemed, in its parallel to the behavior of lower animals, to replicate the structure of organic brains. For all their simplicity, they came closer to the brain than a computer ever could. But of course they had not replicated the brain completely. Walter knew that the robots were not capable of higher thought, of deciding on a particular goal and adapting their actions to achieve it. As Walter concluded, "So-called purposive behavior can be defined in terms of reflexive action without recourse to transcendental teleology."[76] The tortoises had cleared one important step beyond predetermined thought, but their construction suggested that there were more steps to take.

"Lords of the Earth"

This realization helps explain why, while Walter wanted to use his tortoises to break down the mechanical-organic divide, he nonetheless held firmly to another opposition, that between humans and animals.[77] In 1953, Walter presented humanity in the grandest possible terms. As he wrote, "Man, for our present purpose, is specifically what he is by virtue of thought, and owes his survival in the struggle for existence to the development of that supreme function of brain. He is *sapiens*, the thinking species of genus *Homo*—the discerning, discreet and judicious one, even if he does not always live up to all these meanings of the name he has given himself. . . . Other animals have only a glimmer of the light that so shines before men."[78] Walter elaborated on these peculiarly human qualities with respect to fire. Man differed from the other animals because he did not fear it. Confronted with a flame, "he watched it carefully, calculatingly; he accepted the odds and sought to improve them in favour of escaping fresh hurt. He was learning which end of a burning stick he could grasp with impunity."[79] The human control of fire thus pointed to the human ability to think through situations and override natural animal responses.

In a later example, Walter compared a "boy learning to ride a bicycle," with a "foal learning to walk."[80] In the former, there was a "purpose" behind the learning: the boy decided what he wanted to do and controlled his body in a way to execute the desired act. In the latter, Walter identified only a "tendency that is becoming effective."[81] The foal had

> no preconceived ideas about standing up, no image in its brain to give its struggles purpose, no ambition to stand on its own feet like the other fellows. It has a tendency to do so, a tendency to use the feet it was born with in the way provided by its inborn reflexes. It will stumble in its walk and play about until all its motor reflexes are in working order; only the successful ways of running, galloping, jumping, will be remembered; and then, all being well, it will never stumble again.[82]

Humans, Walter continued, could even come to master involuntary functions. Walter referred to "the Yogi"[83] of "Africa and Asia" where "long years are spent in practicing a system of conditioned reflexes whereby the pulse rate, breathing, digestion, sexual function, metabolism, kidney activity, and the like are brought under conscious control."[84] This distinguished humans even from higher animals like the chimpanzee, who were "unable to rehearse the possible consequences of different responses to a stimulus, without any faculty of planning."[85] As Walter put it, "Very early in the human story the brain must have acquired the mechanism of what we recognize in action as imagination, calculation, prediction."[86] To this list of human attributes, he later added another "series of processes: observation, memory, comparison, evaluation, selection."[87]

In a reversal of the standard Orientalist trope, Walter presented this as a higher order of consciousness. The ability for conscious control—in the normal person as well as, in more extreme forms, in the Yogi—could be traced to a "deep physiological division between man and ape," and by implication other animals. Mental function, as we have seen, was linked for Walter to brain waves, which "can be recorded as electrical eddies swirling in subtle patterns through the brain." In animals, these patterns did not exist in the same way;

the electrical "elements of these higher functions" were "only isolated and intermittent."[88] The precondition for these higher brain functions was, according to Walter, homeostasis, the ability to maintain a stable internal state despite changing environmental conditions.[89] True, homeostasis could be found in all warm-blooded animals, but while "for the mammals all, homeostasis was survival; for man, [it was] emancipation."[90] In humans, homeostasis freed the "upper brain," that is the two hemispheres of the cerebrum, "from the menial tasks of the body, the regulating functions being delegated to the lower brain," and this rendered it available for higher-order tasks.[91] Walter cited the Cambridge physiologist Joseph Bancroft and his rendering of Claude Bernard's famous concept of the *milieu intérieur*, which illustrated the principle: "How often have I watched the ripples on the surface of a still lake made by a passing boat, noted their regularity and admired the patterns formed when two such ripple-systems meet . . . *but the lake must be perfectly calm.* . . . To look for high intellectual development in a milieu whose properties have not become stabilised, is to seek . . . ripple-patterns on the surface of the stormy Atlantic."[92]

Human brains for Walter were therefore free realms where brain waves could propagate and be compared. It is clear why under these conditions, brain waves might explain the higher functions Walter named. As we saw, Walter had understood the electrical activity of the visual cortex as a form of line-scanning device that could translate visual sensation into signals. These signals could be transmitted to other brain areas, stored, and recalled at crucial moments in order to be compared with fresh visual signals, allowing the organism to "recognize" familiar objects or patterns. So too, through their manipulation, one could explain the emergence of the world of imagination, a realm detached from sensory reality, which allowed us to aim for unrealized goals, make predictions, and recall memories.[93] Walter even extended his analysis of brain waves to explain personality.[94]

It is true that the precise border between the human and the animal is unclear in Walter's work. At times, he seems to attribute these higher order powers to "association." As he wrote in *The Living Brain*: "For the human brain, even at birth, is so highly organized, the electrical rhythms that sweep it are so suggestive of searching mental activity, that it is difficult amidst so much complexity to tell how soon the ape is left behind. The good fairy's gift of learning by association is found in every cradle."[95] However, in the very next chapter, his account of association draws on Ivan Pavlov's conditional reflex, of which the archetypal case was the salivating dog.[96] So perhaps the difference was one of size and complexity rather than quality. As Walter argued, "From simple animals, through monkeys and apes to Man, the size of the association areas has increased steadily during evolution. We have learnt to master the world by recognizing, remembering and comparing patterns more efficiently than other animals."[97]

In any case, Walter developed a quasi-evolutionary account of the nervous system, where different animals represented different and distinct stages—amoeba, jellyfish, mollusks, fish, birds, reptiles, and mammals possessed nervous systems of rising complexity. At the top of this hierarchy was the human brain. The organic hierarchy was then recapitulated in the inorganic; Walter's robots had an evolutionary order too. W. Ross Ashby's homeostat—or *M. sopora* (the sleeping one), in Walter's nomenclature—which responded to changing conditions through a mechanism of negative feedback in order to maintain a stable temperature was at the bottom of the hierarchy. It belonged, Walter thought, to a different category from that of computers or Wiener's machines; though the homeostat had "a predestined end," it showed "quite unpredictable behaviour in reaching it." It was like a cat or dog dozing by the fireside, which stirs only when disturbed, and then goes back to sleep again.[98] Walter's robots rose to a higher level. The *Machina speculatrix* could be situated above the level of a moth attracted to light, because it was able to find an equilibrium (if it got too close to the light it would turn away). Its behavior was "goal-directed," "unpredictable," and "optimizing." The next level of robot was the *Machina docilis*, which resembled the *Machina speculatrix* except that to the four-element sensory-motor system, Walter added what he called CORA, "conditioned reflex analogue." The technological details of this addition are not relevant to the story here—essentially Walter developed an electronic system that would associate distinct stimuli if they appeared close to each other and repeatedly[99]—but we can see how it seems to integrate a form of pattern recognition that raised the machine up to a new level, certainly to that of Pavlov's dog, and perhaps beyond.

The Mirror

Walter's dual project—the "inverse behaviorism" and a hierarchy of robots mimicking the hierarchy of life—provides the context for his use of the mirror recognition test.[100] When experimenting on the *Machina speculatrix*, Walter noted that it "lingers before a mirror, flickering, twittering and jigging like a clumsy Narcissus."[101] This was "a characteristic dance . . . since it appears *always and only* in this situation" (fig. 3.8).[102]

On the standard "inverse behaviorist" account, one would expect Walter to declare that the tortoise had recognized itself. The *Machina speculatrix* showed

FIGURE 3.8. The tortoise's mirror dance. This drawing depicts the spiraling, exploratory movements of the machine. It demonstrates the stochastic quality of the movement, unlike the idealized movement shown in figure 3.1. W. Grey Walter, "An Imitation of Life," *Scientific American* 182, no. 5 (May 1950): 42–45, on 44.

a similarly idiosyncratic behavior in front of the mirror as self-recognizing creatures. Didn't that mean that the robot's internal mechanism could be assumed to be structurally similar to that at work when other creatures looked in the mirror and recognized their image?

But at the same time, Walter knew exactly what was going on in the "head" of the tortoise, and thus it was clear to him that self-recognition was out of the question. The tortoise had no ability to form and retain a self-concept, let alone apply such a self-concept to its reflection. This is not to say that cybernetic creatures were in principle incapable of recognition. Far from it. One can imagine that higher levels of creatures with more sophisticated apparatuses (at minimum with the CORA device) might be able to apply a self-concept to the mirror image. But in the hierarchy of robots, mimicking the hierarchy of life, Elsie had not achieved the necessary advance.

As compelling as it might be to a naive biologist, Walter knew that the robot's distinct behavior had another explanation. In addition to the sensors and effectors described above, the machine carried a pilot light on its head, which switched off when the driving motor was in full power and when the scanning movement ceased because of a change in light levels. Walter had originally included the pilot light "simply to indicate when the steering-servo was in operation."[103] But when a mirror was placed in the tortoises' environment, the pilot light was swept up into the feedback loop of the scanning sensor. The tortoise approached the mirror in its normal searching spiral. At a certain point, the reflection of its pilot light would activate the light sensor, locking the steering mechanism and thus drawing the tortoise toward the mirror. However, as soon as this occurred, the pilot light would turn off, and so the scanning would begin again. This in turn would reilluminate the light. If by this time the scanning hadn't progressed very far, and so the reflection of the pilot light was still visible, the machine would head back to the mirror, starting the process again. Once the sensor had rotated enough that it no longer could "see" the reilluminated pilot light, however, it would assume again the spiraling motion, until it caught sight of its pilot light once again. Far from an indication that the tortoise had recognized itself, the strange "mirror dance" resulted from the unexpected impact of a pilot light.

Conclusion

The mirror dance was, for Walter, an unsettling experience. At first glance, it seemed a dramatic confirmation of his project. Walter had built his tortoises to show that it was possible to produce a mechanical mind. There was nothing special about the substance of the nervous system; biological brains could be recreated in wires and transistors. That is why Walter's robots could replicate the behavior of animals of various complexities. What greater success then could one imagine than one of the robots passing a test that purported to mark human specificity? The barriers between organic and inorganic seemed to have been torn triumphantly down.

If the tortoise validated Walter's ideas on this front, it challenged them on another. The strong continuities between robots and animals on the one hand, and between robots and humans on the other, were not evidence for Walter that humans were like other animals. Quite the contrary. For the various robots sat on a hierarchy that paralleled and in this way confirmed a hierarchy of organic creation. On this latter hierarchy, humans stood far above other animals. Moreover, the robot hierarchy could not be understood as a quantitative scale, measuring greater processing power, but rather arose through the addition of distinct components. Analogously, the organic hierarchy could not simply be related to an increasing number of nervous connections, or bigger brains. Walter's robots brought him instead to a very similar conclusion to the one that had guided the early baby diarists. One could trace the evolution of thought, showing how each stage built on previous ones, moving from the language of animals to the language of humans. That evolution, however, proceeded in leaps and bounds, as new capabilities were added that were qualitatively different from what had come before. The Oxford Sanskritist Max Müller had distinguished sharply between emotional and rational speech, even if the latter built on the former, because rational speech required a faculty that nonhuman animals simply did not possess. So too Walter distinguished between a machine with alpha waves, which allowed the propagation of signals, from a higher-order machine, which included the capability of storing and associating such signals.

From this perspective, Walter's reference to the mirror test could only be a joke, laying bare the leap of faith involved in the interpretation of mirror behavior, the way one inferred a sense of self from the subject's suggestive movements in front of the glass.[104] For someone who knew what was going on in the head of the tortoise, the idea that it had recognized itself was laughable. The behaviorists and those who followed them in their skepticism about introspection were thus right; the mirror test promised far more than it could deliver. The tension between these two positions shaped Walter's presentation of the mirror dance. He returned to the mirror test again and again but also held it at an ironic distance. He took care not to declare that the tortoise had recognized itself on his own authority. He rather ventriloquized well-meaning but probably mistaken scientists, like the imaginary biologist referred to in the introduction to this chapter.[105]

The surprise caused by *Machina speculatrix* in front of the mirror was not that a robot could recognize itself. It was surprising because the *Machina speculatrix* was constructed to mimic a lower-level animal, one that wasn't supposed to be able to retain and thus recognize its image. Its behavior seemed to demonstrate a capability that Walter knew quite certainly that it did not have. In short, what was disconcerting about the mirror dance was not that *Machina speculatrix* seemed to recognize itself in the mirror as a robot, but rather that it did so as a tortoise.

4

Monkeys, Mirrors, and Me

GORDON GALLUP AND THE STUDY
OF SELF-RECOGNITION

A CHIMPANZEE STARES into a mirror. At first it treats the mirror image as it would treat another animal. It inspects the image with great intensity. It smacks its lips, bobs its head up and down vigorously, and makes what seem like threats. But this is only a passing phase. After a short time, perhaps three or four minutes, the chimp's demeanor changes. Instead of making aggressive gestures, the animal becomes calm and contemplative. It moves its limbs into strange and uncomfortable positions, and then tries to get a glimpse of the result in the mirror. It looks directly at the mirror image and makes exaggerated expressions on its face, sticking out its tongue, again examining the result intently. Now it is playing with its eyebrows, grooming itself in places that it normally cannot quite see.[1]

For the young doctoral student witnessing this scene, Gordon G. Gallup Jr., the transformation was a revelation. In the changing behavior, he discerned a shift from "self-sensation" to "self-perception." He saw, that is, cognition.[2] Whereas at first the chimpanzee perceived only another chimpanzee, perhaps a rival, by the end of the experiment it looked into the mirror and saw itself. The behavior set Gallup on a path to a storied 1970 *Science* paper. In it, he claimed to provide the "first experimental demonstration of a self-concept in subhuman form."[3]

The situation is similar to that of the child in front of the mirror that we encountered in chapter 1. The behavior is highly suggestive; it is almost impossible to resist the interpretation that the chimpanzee had come to recognize its reflection. But, as we have seen, over the course of the first half of the twentieth century, and especially in the English-speaking world, there had been a consistent effort to downplay the broader significance of the mirror test and draw out the complexities of mirror behavior. W. Grey Walter may have wanted to mock the biologist who thought he had discovered signs of self-recognition in his "shelled creature," but in general the difficulties of interpreting mirror behavior had made psychologists cautious about their immediate intuitions. Gallup's confidence that the chimpanzee had recognized itself in the mirror thus seems out of place, a return to the extravagant claims of the late nineteenth century. But Gallup was not naive about the mirror tradition and its difficulties. In fact, he had been trained in the behaviorist tradition and shared its skepticism about introspection, anthropomorphization, and the overinterpretation of experimental data. Nonetheless, like others at the time, Gallup was also wary about absolutizing behaviorist principles and strictly excluding discussion of all internal states. This is why the mirror fascinated him. He thought the mirror—if used in the right way, on chimpanzees but also monkeys and other animals—might be able to open up space within the behaviorist paradigm for the discussion of higher functions, and even concepts like the *self* in nonhumans.

Psychology at Washington State

Gallup's turn to the mirror can be explained by the strengths and limitations of his situation at Washington State University (WSU). Originally founded in 1890 as the Washington Agricultural College and School of Science, WSU received federal funding as a land grant institution and thus was centered on applied subjects, with a focus on those that would serve the practical needs of a rural population.[4] The immediate postwar years were a time of tight budgets and overcrowding for the university, as GIs returned from war and went to college. From 1947 to 1966, the Psychology Department was located in an old wooden dispensary building, brought from the Farragut Naval Training Station in Idaho (fig. 4.1), which it shared with the Education Department. The building had been cut into sections and moved to WSU, where it was reerected on the south edge of the campus. According to a 1958 report, it was "over-crowded, ill-ventilated, noisy, unsafe . . . and dingy."[5] James H. Elder, a faculty member in psychology and chair of the department from 1949 to 1968, was taken by surprise when he first arrived at the university. He had ignored warnings from friends and advisers that "it was just a temporary institution, just a lot of shacks that they've thrown up," and accepted the position as chair without previously visiting the campus.[6] He recalled, however, the "can do" atmosphere: "Most

FIGURE 4.1. Old Washington State University psychology (and education) building, around 1947 (*top*) and demolition in 1974 (*bottom*). WSU Buildings Photographs Collection (Archives 333, box 9, folder 26; box 7 folder 14), Manuscripts, Archives, and Special Collections, Washington State University Libraries. With kind permission from Washington State University Archives.

of us were carpenters of one sort or another and we could just start tearing the building apart if we [could] get approval from buildings and ground for [our] projects."[7] According to Mary J. Kientlze, another professor of psychology, the "dispensary was frequently remodeled by the department faculty—with special accommodations for turtles, mealworms, salmon, photography, and human research tucked into various corners and in hallways."[8]

Faculty and students had to be creative in searching for research space. One professor, F. Dudley Klopfer, located and converted an old truck trailer for his studies in food deprivation in birds, dogs, and pigs.[9] Francis A. Young, who would later advise Gallup's master's thesis, was more ambitious. When the campus post office moved in 1952 and its building became vacant, Young seized the opportunity and acquired the building for the department. He housed a group of about 150 monkeys on two floors.[10] A "can do" attitude was even more necessary for graduate students. Because space in the dispensary, trailer, and old post office was limited, graduate students had to conduct their research "everywhere on campus that could be 'borrowed' for a long enough time."[11]

The situation improved when the Psychology Department moved to new quarters in 1966.[12] Financed by the university with matching funds from federal sources,[13] the department was offered a newly constructed three-story addition to Todd Hall closer to the center of campus: the Johnson Tower. The tower housed offices for the faculty and offered space for human and animal research and facilities for small animals (such as rats), and for teaching laboratories, as well as the Human Relations Center.[14] Large animals and related research moved out to a newly built Comparative Behavior Laboratory, or Primate Research Center, east of the campus, on Airport Road. The building was designed to house a total of about 150 monkeys, 100 birds and 20 to 50 pigs, sheep, goats, deer, and antelopes each—with both indoor and outdoor pens and cages. In addition, it was a breeding facility that served as a source for Japanese and pig-tailed macaque monkeys—the ones that Gallup would use in his student research. The new building was also meant to become a "training center for advanced graduate and postdoctoral students" so it is likely that Gallup spent a good amount of his final two years at WSU there.[15] Despite the upgrade in their accommodations, a spirit of bricolage remained in the department, which was growing quickly at the time.[16] Reminiscing about the new spaces, Elder recalled the "nice shop up there in Johnson Tower." "We need[ed] to make apparatus for working with animals," and for that, it was useful to have a shop right in the basement of their building, with the shop personnel as a helpful source of advice.[17]

Gallup's first interest in mirrors can be placed in this context. It is significant that in the vast majority of his early experiments, the mirror was not used primarily for its reflective properties per se, but rather for a host of other practical reasons. The mirror was a useful and economical piece of equipment when space and budgets were tight. Gallup referred often to these practical advantages. When space was of a premium, Gallup thought, mirrors could simulate social interaction, without needing to put laboratory animals into contact with each other.[18] Animals who were moved into the cage of other animals for social experiments usually required a period of habituation, and they could attack and hurt each other, especially in studies of aggressive behavior.[19] Gallup came to consider that mirrors could be a safer and cheaper alternative: "An environment can be rapidly transformed in terms of incentives to social behavior merely by presenting or withdrawing a mirror."[20]

Of course, such a substitute was justified only if the animals could not recognize themselves in the mirror, and if the mirrors thus sufficiently mimicked social interaction. Gallup noted that though many researchers relied on mirrors in their experiments, no one had sufficiently theorized their use: research on mirrors was, he asserted, "fragmented and isolated," with "no one know[ing] what anyone else is doing."[21] This problem focused Gallup's work up to his PhD, which was an attempt to provide a full account of mirror function, to explain in what circumstances mirrors could be used, and where the peculiarities of mirrors made them a poor substitute for other forms of stimuli. Gallup wanted to analyze rigorously what was in the makeshift world of WSU psychology a de facto necessity: the use of mirrors to simulate social interaction.

Behaviorism and Its Discontents

At first, Gallup saw these peculiarities as a problem, but soon he realized that they might prove valuable in one of the most promising developments within behaviorism, one that had already left its mark on the WSU Psychology Department. When Gallup was there, the overall orientation of the department—like many others at the time—could be described as eclectic neobehaviorism. For the most part, it followed the father figure of American behaviorism, John B. Watson, and his most famous disciple, B. F. Skinner, who promoted a "scientific" study of psychology, meaning one focused wholly on intersubjectively confirmable behavior.[22] The faculty also followed the behaviorist example of conducting the bulk of their psychological research on animals. Behaviorism, by insisting that one did not need to understand what was going on in the black box of the brain, rendered insignificant the differences between animal and human research and thus reinvigorated the methods of an earlier generation of comparative psychologists such as George John Romanes and Charles Darwin.[23]

Elder, who was chair during Gallup's time there, is a case in point. Having earned his undergraduate degree in psychology at the University of Colorado, studying with the behaviorist Karl Muenzinger, Elder moved to Yale for his PhD in 1930, to work with the primatologist Robert M. Yerkes. As he wrote about the time, "Beer, Bethe, von Uexküll, Lloyd Morgan, Jacques Loeb, and Watson stand out as sources of influence in my early years in psychology. As a result of this indoctrination I scrupulously avoided putting thoughts into the head of my rat."[24]

Nevertheless, Elder, like others at Washington, was starting to call into question the behaviorist refusal to draw on mental states to explain behavior.[25] After Yale (he gained his PhD in 1933) Elder had worked at the Yale Laboratories for Primate Biology at Orange Park, Florida, which Yerkes had established in 1930.[26] Yerkes encouraged every laboratory member to produce daily notes to record chimpanzee behavior in the broadest possible way, to "show increasingly other observations than those which relate intimately to our special problems of personal interests."[27] In these notes, Elder figured animal behavior as meaningful, and he frequently framed it in human terms, a direct challenge to the behaviorist prohibition on anthropomorphism. He described the "six immigrants, fresh from Africa," whose interaction produced "a dynamic social situation with gang warfare, sudden and puzzling shifts of alliances, impressive expressions of group cohesion and mutual affection." Elder compared his "efforts at peacemaking" to his parenting practices with his own children, aged seven and ten, and pointed out that he "used the same technique."[28] As Elder concluded, reflecting on his own duties, "I cannot help wondering if we might not have avoided all of our controversy over mind in animals, if Descartes at an early period in his life, had been given the care and feeding of two baby apes."[29]

Elder's questioning of behaviorist principles aligned him with the more programmatic vision of scientists involved in the "cognitive revolution." These scientists reintroduced mental concepts into psychology by assuming that the mind was autonomous, working at least in part independently of stimuli. One of the most vocal attacks on behaviorism came from the young MIT linguist Noam Chomsky. In a scathing 1959 review of Skinner's Verbal Behavior, Chomsky argued that behaviorism was badly suited to account for language.[30] The new interdisciplinary program of cognitive science promised a way to pay attention to entities such as language or mind while holding onto the scientific rigor that distinguished behaviorism. Over the course of the 1960s, behaviorism slowly lost its dominance within psychology.

Elder's example shows, however, that the questioning of behaviorism did not need to come from outside the field; it could also emerge from within the behaviorist ranks, through an engagement with some of behaviorism's most pressing questions and issues. In particular, we can see how, by drawing on the technique of "operationalization," heterodox behaviorists were able to exploit the incomplete overlap between two of behaviorism's most cherished principles: the essentially practical concern to reject introspection in psychological research, in favor of objectively observable events, and the more theoretical concern to exclude the mind.

Operationalism, or "operationism" as it is sometimes called, had its origins outside of psychology. Its principles were first formulated by Harvard physicist P. W. Bridgman in his 1927 *The Logic of Modern Physics*. A concept, to Bridgman, was "nothing more than a set of operations; *the concept is synonymous with the corresponding set of operations.*"[31] Operationalism was a means of turning abstract concepts into reproducible experimental practices, essentially to define them in relationship to the apparatuses used to measure them in an experimental system. Using the example of physical length, Bridgman suggested that it should be defined by the number of times one needed to apply a measuring rod to an object in a straight line. Length could be fully accounted for by an external and observable set of actions.[32]

No other field picked up operationalism with the same fervor as psychology.[33] It is easy to see how the operationalist concern with fully observable and measurable operations would fit the behaviorist concern with fully observable behavior. Indeed, operationalism would allow behaviorism to cleave even closer to its principles, and an operationalism *avant la lettre* can already be seen in Watson's radical behaviorism. When Watson restricted the matter of psychology to observable stimuli and behavior, he seemed to exclude from discussion not only the mind, but also a number of internal and yet non-mental states such as hunger and sexual drives, what Watson called "internal stimuli."[34] The existence of such stimuli could not, however, be plausibly denied, and Watson included them in his framework. As he wrote, "By stimulus we mean any object in the general environment or any change in the tissues themselves due to the physiological condition of the animal, such as the change we get when we keep an animal from sex activity, when we keep it from feeding, when we keep it from building a nest."[35] To get around this problem, as the quote shows, Watson defined these "internal stimuli" in purely observable terms: changes in the animal's environment and treatment.

Less than a decade after Bridgman's publication, the psychologist S. S. Stevens published a pair of articles announcing that "the course for psychology is clear": Psychologists needed to integrate operationalist criteria into their work. To Stevens, this was the only way for psychology to be "fortified against meaningless concepts."[36] Benton J. Underwood—in the 1966 edition of his textbook *Experimental Psychology*, the most widely used text for undergraduate teaching in psychology, dominating university curricula all over the country[37]—remarked that "most experimentalists in all disciplines tacitly or actively accept operational definitions as a means of specifying the empirical basis of a discipline. That is, they use operational definitions in order to define the phenomena of nature with which a discipline concerns itself."[38] Underwood opposed an operational definition to literary or dictionary ones: it "identifies a phenomenon by specifying the procedures or operations used to measure it."[39] He admitted that the psychologist was often interested in concepts that were known anecdotally from his own observations or even from novelists, and he accepted that operationalizing them might result in some key elements of their meaning being lost. Underwood argued, however, that this was a necessary sacrifice.

Operationalism was particularly valuable because it offered an opening for internal physiological states, which had previously been banished from discussion. Indeed it is because of this that operationalism persisted in a psychology oriented around behaviorism, even though operationalism had come under consistent criticism from philosophers of science, who by midcentury had come to deplore its naive realism.[40] Taking their cue from logical positivism, and especially the Vienna Circle,[41] neobehaviorists suggested that by "operationalizing" concepts such as *drive* they could bypass a reliance on introspection and thus integrate such concepts into behaviorist analysis.

The debate over drives revolved around the concept of reinforcement: the way in which behavior could be encouraged by the use of rewards in an experimental system. The existence of a drive, which could be reduced by some reward (e.g., reducing hunger with food), would explain why behaviors that led to that reward would be preferred and repeated.[42] Watson disliked talk of reinforcement because it threatened to import human understandings of what the animal wanted. The idea that a particular action (e.g., giving the animal food) would encourage the repetition of behavior relied too heavily on unverifiable assumptions about the animal's desires.[43] But for neobehaviorists such as Clark L. Hull, reinforcement provided a valuable operationalist account of such drives.

Hull formulated the "'law' of *primary reinforcement*" thus: "*Whenever a reaction (R) takes place in temporal contiguity with an afferent receptor impulse ($\dot{s}$) resulting from the impact upon a receptor of a stimulus energy ($\dot{S}$), and this conjunction is followed closely by the diminution in a need (and the associated diminution in the drive, D, and in the drive receptor discharge, S_D), there will result an increment, Δ ($\dot{s} \rightarrow R$), in the tendency for that stimulus on subsequent occasions to evoke that reaction.*"[44] The crucial element here is the Δ ($\dot{s} \rightarrow R$). If upon repetition of an experiment a particular stimulus (S) would produce a response of *different* magnitude or with a *different* latency time, then this could no longer be explained by Watson's simple stimulus-response system, which implied a stable relationship between the two. It required, for Hull, the appeal to a drive—that is, a variable placed in between Watson's stimulus and response—or, in his own terminology, between "receptor-effector connections."[45]

Hull took as his example a rat, on one side of a barrier, exposed to electric shock on its feet. The rat leaped over a barrier to escape, and on repetition of the experiment did so more quickly. At the same time, "futile reactions," such as leaping against the container walls, biting the floor, and squeaking, decreased.[46] That is, the electric shock (S) had reinforced the barrier leaping (R). The rat's behavior within this experimental situation was fully observable, and experimentally repeatable. But most importantly for Hull, the changing relationship between stimulus and response showed that the "the termination of the shock effect" was a "need" connected to some internal drive.[47]

The neobehaviorist Abram Amsel picked up Hull's treatment of drives in the 1950s and 1960s to develop his theory of frustration.[48] As Amsel and his colleague Jacqueline Roussel defined it, frustration was "a state which results from the nonreinforcement of an instrumental response which previously was *consistently* reinforced." The basic insight of this theory was that if drives existed, their frustration (e.g., refusal of food) would lead to an "increase in generalized drive strength" and thus an observable increase in the conditioned response. Again we have an operationalized account: the changing relationship between stimulus and response in repetitions of an experiment pointed toward a mediating drive. In Amsel and Roussel's words: "In order to show that frustration is a motivating condition it would seem necessary to demonstrate that its presence increases some aspect of behavior."[49] Amsel then demonstrated how rats ran faster along a track when in an earlier trial the expected food reward had been removed.

Neobehaviorists like Hull or Amsel did not push the boundaries of behaviorism very far. Drives were sufficiently simple and tied into biological processes that they did not consider it a stretch to infer their existence in animals.[50] But their work was important because it showed how operationalism could be used to bend the rules of behaviorism, how by shifting attention from the animal to the way in which responses to the same stimuli might change under precisely controlled conditions, behaviorists could talk about internal states. Once pushed back, the boundaries showed themselves to be quite elastic. Some, including the heterodox neobehaviorist Edward Tolman, took operationalism as a license to widen his perspective beyond simple drives; he provided a redefinition of the psychological concept of *demand*, which, as for Hull, acted as an "intervening variable" between stimulus and response.[51] Writing in the logical positivist journal *Erkenntnis*, Tolman defined *demand* broadly as "desire, wish, purpose, want, need, value, motive."[52] As examples he offered appetitive and aversive "primary demands," such as food demand and object-absence demand (e.g., of a pricking or cutting object), but also higher-order, "secondary demands," such as demands "for particular styles of wearing apparel" or "for the accomplishment of particular social goals."[53] In the 1950s, primatologist Harry Harlow had used a similar approach to operationalize curiosity, to explain why primates worked with puzzles even in the absence of food rewards.[54]

Gallup's Operationalism

Gallup engaged with these ideas from his earliest work. Operationalism allowed him to combine the two interests that had marked the beginning of his career. It offered the possibility of transforming the patient analysis of how mirrors functioned in experiments into the more ambitious project of integrating higher order concepts into behaviorism. One of his earliest experiments, published even before he was awarded his master's degree, built on Amsel's research on reinforcement and frustration. He placed food at the end of a straight alley as reward for rats who ran its length.[55] Gallup expanded on Amsel, however, by registering aggression, drawing on the recent "frustration-aggression" hypothesis, which suggested that the former could elicit the latter: if partial reinforcement led to frustration (Amsel), and frustration led to aggression, would partial reinforcement lead to aggression? Gallup's experimental results suggested that it did.

The mirror served as a valuable addition to Gallup's experiments on the "frustration-aggression" hypothesis. Gallup used the same setup as in the rat experiment described above, but this time at the end of the alley rather than food, the rats found a mirror, which they were allowed to inspect for thirty seconds. In Gallup's master's thesis from 1966, he explained why he thought the mirror might be a good reinforcer. Research by R. A. Butler from the mid-1950s had shown that monkeys responded better to the sight of other monkeys rather than other species like dogs or objects such as toy trains. In addition, if a stimulus was in motion and changing, it was more effective than when stationary. Mirrors fulfilled both of these properties. In fact, on the latter point they might even be more effective than conspecifics; Gallup noted that because the mirror provided instant replication of all actions and gave the animal control over them, a mirror image was able to "maintain stimulus change at a high level" and thus could "reinforce responsiveness."[56]

If the mirror did function as a reinforcer, though, it would be a great coup. Because the mirror did not respond to any obvious biological need, it would point beyond the physiological drives that were "more widely accepted" by behaviorists to the "so-called intrinsic motives, unrewarded approach behaviors, and curiosity-investigative incentives."[57] That is, the mirror promised a means to operationalize higher-order internal functions. As it turned out, the rats appeared indifferent to the mirror, and the research was never published.[58] Nonetheless, the experiment set the trend for his future studies. Gallup wanted to build on earlier work on the drives and use the mirror to open up internal states of even greater complexity.

While in the themes of his studies Gallup seemed to be moving ever further away from the behaviorist orthodoxy, in his experimental practice he cleaved closely to it. It is significant that while embracing the "so-called intrinsic motivations" as a topic, he distanced himself from the language used to describe them. In the body of his work up until 1968, Gallup studiously refused any reference on his own account to internal states. When referring to the use of drives, he tended to use noncommittal language: for instance, "the efficacy of a reinforcer is often assumed to be a function of . . . the strength of an appropriate underlying drive."[59] Because the field of motivation studies lacked terminological unity, Gallup argued, he attempted to operationalize thoroughly all the terms. Already in his rat aggression experiment, he used the "Klein-Hall scale" to measure aggression.[60] Later he suggested the pupillography technique developed by E. H. Hess and J. M. Polt in 1960 for measuring emotional states in primates.[61] It is fitting then that later, in a review of a book by the philosopher Bernard Rollin on animal awareness, Gallup described himself as a "logical positivist."[62]

The "intrinsic motivations" elicited by the mirror turned out to be well suited to this approach; they were more susceptible to operationalization than those referring "to internal stimuli or drives serving homeostatic or survival functions," because unlike hunger they did not depend on unverifiable internal physiological states. Ironically, Gallup thought that in contradistinction to the drives, these "*intrinsic* motivations" could be refigured so as to be based on "*external* sources of motivation," producing "exteroceptively aroused" behavior.[63]

Gallup's attempts to operationalize these intrinsic motivations led him to develop a taxonomy of behaviors in front of the mirror, that is, an accounting of its functioning as a piece of experimental equipment. For his master's thesis, Gallup built a wooden box that could be fitted over the door of a metal monkey cage (fig. 4.2). Once the box was in place and a guillotine door raised, the monkey was confronted with a door attached to a low-tension spring, which it could open itself. Behind that door the monkey found either a plain wooden board or a mirror. Gallup was able to use six preadolescent and experimentally naive monkeys (one male and three female pigtailed monkeys, one male Japanese monkey, and one male rhesus monkey), with the support of a student grant from the NIH.[64] Testing each animal in turn, Gallup found that the monkeys kept the door open for longer when it gave access to a mirror. This, he argued, confirmed the mirror's reinforcing properties.[65]

For his PhD thesis, Gallup expanded his analyses of the mirror beyond its effects as a reinforcer. As he argued, "There is no question but that a mirror is an exceedingly complex stimulus," and "psychology is without a heuristic and unified conceptual framework to interpret the behavior of an organism in the presence of its mirror image."[66] He thus set himself the task of producing such a "conceptual scheme," which in Watsonian language would "integrate and provide a framework for making predictions as to the possible outcome of mirror stimulation under various conditions."[67] Most important for our purposes, Gallup now "dimensionalize[d]" the responses to mirror-image stimulation "into other-directed and self-directed behavior."[68] This analysis informed the structure of the dissertation, which moved from the simplest "motivational properties" of the mirror (chapters 4, 5), through its effect as a social stimulus (chapter 6), and finally to its function when the image was recognized as a reflection (chapter 8).

Again, we should stress that Gallup remained true to his commitment to behaviorist experimental practice. Though using the language of the *self*, his descriptions remained purely at the level of behavior, noting the distinction

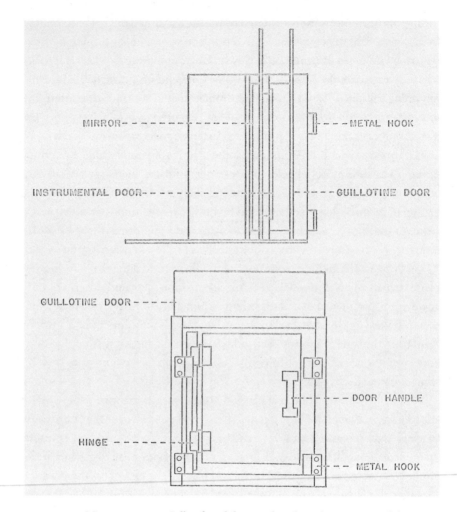

FIGURE 4.2. Mirror apparatus. Gallup found that monkeys kept the instrumental door open longer when it gave them access to a mirror, thus demonstrating the mirror's reinforcing properties. Gordon G. Gallup Jr., "A Technique for Assessing the Motivational Properties of Self-Image Reinforcement in Monkeys" (master's thesis, Washington State University, 1966). Reprinted with kind permission of Gordon G. Gallup.

between, say, aggressive displays toward the mirror image (as if to a conspecific) and self-grooming in front of it. Indeed, he refused to include "references to the self" in his definition of mirror action.[69] While in the master's thesis he had used the term "self-image stimulation," in the PhD, wary of the cognitivist resonances of that formulation, he preferred the more neutral "mirror-image stimulation."

Nevertheless, the identification of a range of different behaviors in front of the mirror at the very least suggested the existence of different mental states. In the seventh chapter, Gallup described what seemed to be the process whereby some animals came to recognize the mirror image as their reflection. For many the mirror acted as a social stimulus; the animal saw another animal in the mirror and then reacted to it as it would to a conspecific. But Gallup noted that because of its direct mimicry of the animal's actions, the mirror image constituted an "atypical social response." More specifically, the "supposed interactive S-R chain of events is upset. In front of a mirror, the animal creates its own stimulus situation and only responses on the part of the animal itself serve as stimuli for additional responding."[70] While the monkey first seemed to identify the mirror image as a conspecific, the peculiarities of the interaction could lead the animal to place that judgment into doubt.

How animals responded to the strangeness of the mirror image varied. For some species it did not matter, and "the mirror image appears to retain enough social stimulus characteristics to perpetuate other-directed-responding in these species."[71] Most primates, in contrast, would simply lose interest in the mirror. Their social response would "extinguish."[72] In his PhD thesis, Gallup witnessed a very different reaction in a single adolescent male chimpanzee, which I recounted at the beginning of this chapter. In response Gallup allowed himself to break the behaviorist prohibition on invoking internal states and appealed to the concept of *recognition*: "In front of a mirror an individual is confronted with a dualistic perception of self. Whether this dualism is recognized as such should determine the degree of other-directed versus self-directed behavior which ensues."[73]

The PhD thesis discussed this self-directed behavior only briefly. It arose first from the observations of the behavior of a single animal, and the longer treatment of self-directed behavior in the final chapter was predominantly a discussion of the work of other scientists on humans. Nevertheless, important here is that Gallup had identified a shift in behavior in front of the mirror, just as Amsel and Hull had found a shift in response to certain stimuli. And already Gallup was willing to use nonbehaviorist language, however tentatively, to explain it. That is, the unstable stimulus properties of the mirror as a piece of experimental equipment suggested a way to operationalize the self-concept. Moreover, this part of Gallup's work pointed to future research. The last part of the penultimate chapter was a list of possible different experiments to test for self-directed behavior in primates, including the first outline of what would become his career-making "mark-test."[74]

Chimpanzee Self-Recognition

Gallup graduated from Washington State in 1968 and accepted a tenure-track position as an assistant professor of psychology at Tulane University the same year. The new post offered him the opportunity to conduct his animal research on a large scale, especially through Gallup's new affiliation with the prestigious Delta Regional Primate Research Center in Covington, Louisiana, about an hour's drive from campus. The center had been opened in 1964, as one of eight in the United States that made up the National Primate Research Center Program funded by the NIH.[75]

In the summer after his first year of teaching, Gallup was a visiting scientist at Delta Regional and conducted the study there that would make his name. The paper was published in *Science* in 1970: "Chimpanzees: Self-Recognition." In the paper, Gallup claimed that chimpanzees recognized themselves in mirrors and thus that they had a "self-concept."[76] Four feral preadolescent chimpanzees, two males and two females, were presented with a full-length mirror placed at a distance of 3.5 meters in front of their cage. They were housed individually for the duration of the experiment and had previously been isolated for four days, in individual cages in the corners of an otherwise empty room. After two days of eight-hour habituation, the mirrors were moved closer to the cage, at a distance of 0.6 meters from the cage, and kept there for eight days. Through a hole in the wall, two experimenters observed the animals' behavior during two fifteen-minute sessions every day, noting down any behavior at least every thirty seconds. At first, the animals responded socially to their reflected images, showing behavior such as vocalizing, bobbing, and threatening. This kind of behavior, however, declined quickly. Instead, the chimpanzees began to exhibit self-directed behavior in front of their reflections: grooming parts of the body that were otherwise invisible to them, picking material from their noses while looking into the mirror, and so on (fig. 4.3).

In order to demonstrate experimentally that self-recognition was present in the animals, Gallup removed them from their cages after ten days of mirror exposure and applied a red alcohol-soluble dye to an eyebrow ridge and the opposite ear of the anaesthetized animals (fig. 4.3). After recovery, the animals, which had in the meantime been returned to their cages, were placed in front of the mirror for thirty minutes. Gallup recorded a high number of mark-directed responses, which included inspection of the fingers after touching the marked area. To control the experiment, he performed the mark test in two additional feral chimpanzees with no prior exposure to mirrors and found that

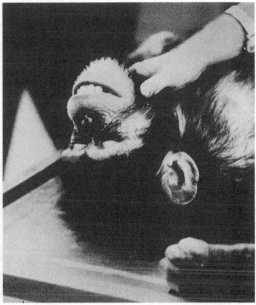

FIGURE 4.3. *Left*: A chimpanzee picking material from his nose while looking at himself in a mirror placed outside the cage (self-directed behavior); *right*: anaesthetized chimpanzee during marking procedure. Gordon G. Gallup Jr., "Chimpanzees and Self-Awareness," in *Species Identity and Attachment: A Phylogenetic Evaluation*, ed. M. A. Roy (New York: Garland, 1980), 223–43, on 228 and 230. Reprinted with kind permission of Gordon G. Gallup.

they failed to show any mark-directed response. From this, Gallup concluded that self-recognition in the original set of chimpanzees had been learned at some point during the ten days of mirror exposure.

The key to the 1970 paper was the mark test, which as we saw had been developed independently at about the same time by the child psychologist Beulah Amsterdam for human infants. For Gallup this demonstrated experimentally that chimpanzees had a self-concept. From a behaviorist standpoint, however, the argument was weak. The animals' behavior in front of the mirror was certainly suggestive, but the whole point of behaviorism was to prohibit the attribution of mental states (such as that of self-recognition) to animals on such a basis. In 1981, the neobehaviorist B. F. Skinner, along with his colleagues Robert Epstein and Robert Lanza, published an article in *Science* provocatively entitled "'Self-Awareness' in the Pigeon." In direct response to Gallup's paper, the researchers set out to show that Gallup's mark test was insufficient to allow him to invoke the self-concept. Epstein, Lanza, and Skinner showed that pigeons

FIGURE 4.4. "Self-awareness" in the pigeon. The pigeon was taught to peck at a dot on its chest visible only through the mirror. Robert Epstein, Robert P. Lanza, and B. F. Skinner, "'Self-Awareness' in the Pigeon," *Science* 212 (1981): 695–96, on 696. Reprinted with permission from AAAS.

could be trained to exhibit seemingly self-aware behavior. And because no one claimed that pigeons had a self-concept, that behavior was to be attributed entirely to "environmental factors," that is, a "nonmentalistic account."[77]

In the experiment, the researchers placed the pigeon into a classic Skinner box (32 × 36 × 42 cm), but equipped with a mirror (34 × 21 cm) behind a plexiglass wall on its right side (fig. 4.4). The researchers established two "repertoires" in the animals to produce the behavior in front of the mirror that Gallup and others had attributed to a self-concept. First, while the mirror was covered, they taught the bird to look for blue dots on its own body and to peck for them using food reinforcement (when the animal pecked at itself between one and five times, it was offered food through a dispenser). Second, they taught the animals to use the mirror. The animal received food for looking into a mirror and also when turning around to peck a spot on the wall behind it where a blue dot had been flashed. After establishing those two repertoires, the researchers conducted their test. They placed a blue dot on the pigeon's breast, which was visible to an observer when the pigeon stood upright. For the pigeon, however, the dot was invisible because it was covered by a bib around the animal's neck. It became visible to the pigeon when it looked at itself in a mirror. The result of this preparation was that in front of the mirror the pigeon began to peck at the place on its breast under the bib that corresponded to the dot.

The experiment was part of a larger project of debunking nonbehaviorist explanations of complex behavior. The year the *Science* article appeared, Epstein completed his doctoral dissertation in psychology with Skinner at Harvard.[78] A year later, the research formed the basis of an educational film, *Cognition, Creativity, and Behavior: The Columban Simulations*.[79] There, Epstein and his former adviser produced various types of behavior that looked like the results of cognitive processes, such as language and problem solving, in addition to self-recognition, and then showed instead that they could be explained by behaviorist learning. In the film, Epstein, who had since taken the position of executive director of the Cambridge Center for Behavioral Studies,[80] not only made it very clear that he didn't think that something like the self in animals existed; he also suggested that "this kind of concept . . . just obscures the search for the controlling variables of the behavior it is said to produce. That is, it doesn't tell you what the real causes are and it keeps you from looking [for] them." To Epstein, Gallup's explanation was circular in that it presupposed self-recognition in animals and then used the concept to explain their behavior.[81]

In a number of papers written during the 1980s, Gallup responded to Epstein's rebuttal. As he wrote, "If I were to teach a pigeon to peck the correct alternatives on an answer sheet to the Graduate Record Examination, and as a consequence it attained a combined verbal and quantitative score of 1500, would that provide a plausible account of the performance of college seniors taking this test?"[82] Nevertheless, Gallup seemed already in 1970 to have recognized the weaknesses of the mark test standing on its own; it did not live up to the standards required for a truly operationalized concept. For over the subsequent five years, he developed a program of mirror research that resembled more the slow and detailed analysis of the 1968 PhD thesis than the cognitivist pronouncements of the 1970 paper. As he put it in a 1975 synthesis (the first of a series of long synthetic papers that drew on the entirety of his work with mirrors), he wanted to develop an "operational definition of self-awareness," and this required more than the animal scratching a colored mark suggestively.[83]

After the Mark Test: Understanding Mirrors

The creation of such an operational concept involved a two-step analysis of the mirror as a piece of experimental equipment. First, Gallup analyzed the peculiarity of the mirror as a social stimulus: how responses to it differed from those when an animal was presented with a conspecific. That is, he would show that the mirror was a stimulus sui generis. Second, he suggested that for some animals, the stimulus properties of the mirror could change over the course of the experiment. That change, Gallup argued, could not be explained by behavior or prior conditioning; one was compelled to invoke a self-concept. The operationalized self-concept depended on the close analysis of the different ways in which a mirror functioned in animal experimentation and its unstable nature as a stimulus.[84]

Throughout the early 1970s, Gallup engaged in mirror experiments on animals for whom there was no suggestion that they had a self-concept, especially birds and fish.[85] The goal here was to test the peculiar properties of mirror stimulation, with respect to social stimulation, by bracketing the larger and more contentious question of self-awareness. In these cases the mirror was presented as a *supernormal social stimulus*, a term borrowed from ethology designating an exaggerated version of a normal stimulus, a kind of hyperreality, which led to a preference of the animal for it over the naturally occurring stimulus.[86]

As the very term suggests, however, *supernormal stimuli* were ambiguous: Was the stimulus super*normal* in the sense that it was a heightened form of normality? Or was it *super*normal, that is, exceptional? That is, did mirrors function as good social stimuli? Or rather were they so different that they no longer produced usual social interactions? These questions were at the fore in a 1971 study of goldfish. Gallup and his colleague at Tulane University John Hess wanted to challenge the idea that mirrors could be used to produce an aggressive display (like an encounter with a conspecific), as researchers had suggested for male Siamese fighting fish.[87] In contrast Gallup argued that the attraction of the mirror, its "appetitive properties," were attributable to "the mirror reflection per se."[88] To show this, Gallup conducted his experiment on goldfish, a species that did not usually show any aggressive displays. For the experiment, he used ten experimentally naive goldfish from a local pet store, kept on a diet of tubifex worms. The investigators created an underwater alley of 0.12 meters width and 0.51 meters length using plexiglass partitions in a ten-gallon aquarium. At one end of the partition, they placed a 0.12-by-0.12-meter glass mirror and a small container housing a target fish on opposite sides. After a brief period of adaptation to the experimental surroundings, the fish underwent preference testing for 60 minutes, with three 10-minute observations after the 10, 30, and 50 minutes in which they recorded the time the fish spent in front of either the mirror or the target fish. Gallup and Hess found that a considerably greater amount of time (1,262.5 seconds as opposed to 402.9 seconds) was spent in front of the mirror rather than the target fish, which constituted a "distinct preference for mirror-image stimulation." To the researchers, this was due not to aggression (an explanation where the mirror as a stimulus would have been replaceable by a conspecific showing an aggressive display) but to the fact that mirrors constituted an "atypical form of social stimulation." The animal had had no previous social encounters that resembled the behavior of its mirror image; the mirror was preferable because of its novelty. Overall, Gallup warned, the effects of mirrors "may impose limitations on the use of mirrors in animal social experimentation."[89] A year later, Gallup pronounced his warning even more forcefully. Although mirrors had "distinct social stimulus properties for some animals," based on his findings, he argued "against the advisability of using mirrors to simulate *normal* social encounters."[90] Supernormality had become nonnormality.

The mirror thus represented a unique type of stimulus, even for animals that did not seem to recognize the image to be of their own bodies. If Gallup could show how for some animals the stimulus properties of the mirror changed over time, just like Hull had shown that the electrical shock produced barrier leaping more quickly when the experiment was repeated, he would have an "operationalized" account of self-awareness. In his 1968 doctoral dissertation, Gallup had suggested that there was not a qualitative but rather a quantitative difference between animals that exhibited other-directed and those that exhibited self-directed behavior in front of the mirror. The quantitative factor was time.[91] The animals needed sufficient time (which depended on their species) to recognize the abnormality of the mirror as a social stimulus and come to see the image as a reflection of their own bodies. Even in the 1970 paper, Gallup had measured the frequency of other- and self-directed responses over time. The graphs showed that by the third day, other-directed had been replaced by self-directed responses; the chimpanzee had recognized itself (fig. 4.5).

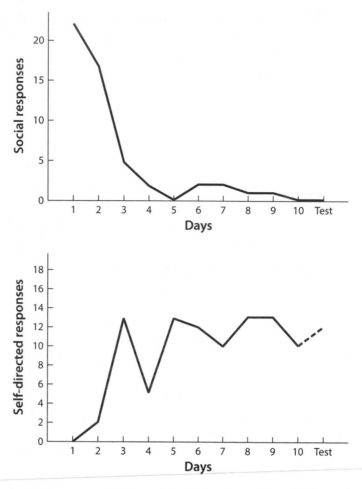

FIGURE 4.5. "Chimpanzees: Self-Recognition." Other-directed (social) behavior (*top*) was replaced by self-directed behavior (*bottom*) by day 3, which Gallup interpreted as a sign of self-recognition. Gordon G. Gallup Jr., "Chimpanzees: Self-Recognition," *Science* 167 (1970), 86–87, on 86. Reprinted with permission from AAAS.

Gallup also tested self- and other-directed behavior over time in adult stump-tailed macaques and rhesus monkeys and then in preadolescent cynomolgus monkeys. In those cases, no shift was visible.[92] Over the next few years, Gallup would confirm this result, again placing emphasis not simply on one particular set of actions performed in front of the mirror but on the way the mirror elicited changing behavior, a transformation of its stimulus properties.

Gallup's operationalized self-concept required more than just the mark test; that test needed to be integrated into a far larger matrix of behaviors in front of the mirror. The course of Gallup's experiments from 1970s showed a pattern of changing behavior in certain animals when they were confronted with the mirror. Primates with a self-concept would all address the mirror image as another animal (other-directed behavior), but over time other-directed would be replaced by self-directed acts. Moreover, these primates would become dramatically more interested in their mirror image if given sufficient cause (the mark test).

The results of these different experiments could not simply be explained by a simple operant account, as in the Epstein simulation. Epstein's refutation of Gallup relied on the reproduction of a single behavior in response to a well-defined stimulus (the mirror), abstracted from the experiences beforehand (Gallup had insisted that he *hadn't* trained his animals to behave in that way). But Gallup's operationalized self-concept, like Hull's and Amsel's operational-ized drives, was no longer based on the one-off test. For the mirror stimulus produced a different behavior at the end of the experiment from the behavior it had displayed at the beginning (broadly other to self-directed). And for that reason, Gallup was compelled to rely on more than conditioning to explain it. When Gallup wrote about the animal's "ability to discern correctly the identity of the reflection," we should read the verb (*to discern*) in its most active and transformative sense, the movement from nonrecognition to recognition.[93] And this suggested that an integrated self-concept "may be much more cogni-tive than mechanistic."[94]

Conclusion

In this chapter, we have seen how Gallup sought to introduce a study of higher cognitive processes into a behaviorist framework. He did this through paying close attention to the workings of the mirror in psychological experiments, examining the type of behaviors it could elicit and how it differed from other stimuli, like the presence of conspecifics or other reinforcers. And, because in certain situations he saw the mirror's stimulus function change—in different iterations of the same experiment on the same animal it could elicit different behaviors—he felt compelled to infer an intermediate variable between stimu-lus and response: the self-concept. In his own words, Gallup had "operational-ized" the concept of self and thus given a response to the central question of mirror behavior: How could we be sure that a nonlinguistic creature had truly recognized itself? At the same time he had challenged one of the remaining bastions of human specificity: he had demonstrated that nonhuman animals like the chimpanzee had a self-concept too.[95]

After Gallup, we see study after study that used variations on the mark test to extend the realm of self-recognition even further into the animal kingdom. Dolphins, elephants, dogs, horses, magpies, manta rays, cichlids, squid, and even ants were placed in front of a mirror to see if they recognized themselves.[96] But in the vast majority of these cases, the motivation for research was diametrically opposed to that which had recommended the use of mirrors for Preyer. Rather than marking the boundary between humans and animals, now the mirror test served as one of the most effective ways of breaking that boundary down. What is more, it has been broken down in a way that recognized cognitive abilities in nonhuman animals that had previously been denied in them and thus had tended to raise them up to the level of humans rather than, as in many other works on the animal-human difference, to emphasize our own proximity to animals. It is worth recalling that Gallup's *Science* paper was published just five years prior to the publication of Peter Singer's *Animal Liberation*, a founding document of the animal rights movement.[97] The fact that animals showed signs of mirror self-recognition became a central motive for scientists to participate in animal activism. The psychologist Diana Reiss, in her 2011 *The Dolphin in the Mirror: Exploring Dolphin Minds and Saving Dolphin Lives*, argued that her insights into the intelligence of dolphins made them ever worthier of our protection.[98] We've come a long way, then, from the child in the mirror to "Save the Whales."[99]

Gallup's work provided the impetus for another tradition of mirror testing, one that, while remaining embedded in the humanism long associated with the mirror-testing tradition, had come to flip its meaning. To see why, we will now turn to the work of the French child psychologist René Zazzo, who spent a lifetime researching twins and mirrors. Zazzo is uniquely qualified to act as a bridge between the two parts of this book. He engaged with all aspects of the mirror-test tradition discussed thus far, from Darwin to Gallup, foregrounding the ambiguity of behavior and the problem of interpretation. The mark test, however, allowed Zazzo to redirect his attention to other aspects of the mirror encounter, emphasizing the strangeness and alterity of our reflections. Perhaps mirror recognition should not be seen as a developmental success or a sign of advanced cognitive ability. The image was an illusion, and identification with it, at some level, an error.

Interlude

THE FIRST ONE HUNDRED YEARS of the mirror-test tradition were dominated by a simple and seemingly intractable problem. Given that mirror self-
recognition had displaced the language demarcator, the test had been used
predominantly on subjects who were unable to verbalize their inner world:
human infants and nonhuman animals. The question arose how one could be
sure that the seeming signs of self-recognition in the subject's behavior were in
fact that. This problem motivated the most important early criticisms of the
test and, in the postwar period, drove a range of proliferations, including
the use of different media in order to build a firmer foundation for the mirror
researchers' conclusions. It also explains why Gallup's and Amsterdam's mark
tests were received across Europe and America with such enthusiasm.

No person better exemplifies the impact of the mark test on the self-
recognition tradition than René Zazzo (1910–95). His work brings together
the threads I have discussed so far. In January 1945 after the birth of his son
Jean-Fabien, Zazzo followed Darwin and Preyer in deciding to compose his
own baby diary, which included a moment of mirror self-recognition:

> February 17, 1947—My mother holds Jean-Fabien in her arms. She places
> him in front of the mirror over the chimney. I ask: "Who is this?" and as he
> does not respond, his grandmother repeats the question. He gives an em-
> barrassed smile, then turns his head away so that he does not see his image
> anymore. His grand-mother turns him around half-way, and he finds him-
> self nose-to-nose with his image. Contortions with the same confused
> smile. Sudden redness of his face. Finally, he pulls himself together and
> leaves the mirror with the gesture and farewell phase "*ava bébé séhi*" (*au
> revoir bébé chéri* [goodbye, sweet baby]). That's the first time that I observe
> this reaction of disarray and avoidance in front of the mirror.[1]

Jean-Fabien had recognized the reflection significantly later than the children of other figures we have encountered—just over two years in comparison to around sixteen months for Preyer—and we might chalk this up to his unusually slow development.[2] But as we will see, the delay was rather due to his father's experimental methods, which followed the norms of the mental testers in the United States.

In 1933, after studies at the Sorbonne in psychology and philosophy, Zazzo had traveled to the American East Coast, working with Arnold Gesell at the Yale Child Study Center and Lois Meek Stolz at the Child Development Institute, part of Teachers College, Columbia University. The experience proved influential. Later he wrote that before his trip "I did not know how to see. My observations sinned not only through imprecision, but more seriously through blindness or invention, interpretation."[3] Following Gesell, Zazzo became increasingly reliant on video in his research, using it to tease apart stages of child development, and he became suspicious of what he thought was an uncritical use of the mirror self-recognition test, which had justified the earlier dating.

For Zazzo, mirror research was "a domain where the ideas are so much richer than the facts," and for that reason "an attitude of prudence" seemed warranted.[4] The predominance of speculation over evidence explains why he came to reject the dating of several predecessors, most notably Darwin and Preyer (who placed mirror recognition eight and fourteen months respectively); they read into the mirror encounter far more than the child's behavior justified. Zazzo deferred rather to the "psychometricians," including Gesell and Stutsman, who had demonstrated that children came to recognize themselves in the mirror only around their second birthday or even later.[5] As Zazzo summed up, "The dating became later and later as the observation became more rigorous and the criteria more exigent."[6]

Despite embracing such rigor, he worried that it could be taken too far. Zazzo's qualms can be seen in his reaction to another figure we have met in this book: William Grey Walter. On returning from America to France, Zazzo gained a post at the Laboratoire de psychobiologie de l'enfant, part of the CNRS, the large and prestigious state research organization. Zazzo would stay there for his entire working life, becoming director in 1950.[7] Thanks to his position, Zazzo participated in the Study Group on the Psychobiological Development of the Child, which held four annual conferences between 1953 and 1956. The international and interdisciplinary group included Walter as well as the psychoanalyst John Bowlby, the ethologist Konrad Lorenz, the anthropologist Margaret Mead, and the cognitive psychologist Jean Piaget.[8]

The meetings clearly left an impression. As we saw in chapter 3, when confronted with a mirror, Walter's robot displayed a form of behavior that could easily be taken for self-recognition, despite its clear inability to recognize itself. When Walter presented his findings to the group, Zazzo was impressed by the "autonomy" of Walter's machine, the way it broke free from predetermined courses of action, and thus showed that "we have not yet finished exhausting the resources of the mechanism, nor the richness of a description of the *object* as such." The machine was able to reproduce many of the characteristics researchers had previously attributed to the "subject." This even seemed to extend to self-recognition, "consciousness in its most specifically human activity," and here in their conversation Walter compared the robot's mirror behavior to that of Zazzo's son, quoted above.[9]

Zazzo worried nevertheless that the argument collapsed into absurdity. He recognized Walter's comparison as a joke: "It was obviously for Grey Walter a point of humor, but that kind of strategic humor that allowed him to exceed, without too much scandal and without excessive engagement, the permitted frontiers of speculative daring." Zazzo was not taken in. In his view, in order to prevent their machines from becoming "the totems of a new magic," cyberneticians needed to push a "mechanistic explanation" as far as possible, but thereby also recognize its limits: "the imitation of life is not life itself."[10] In particular, they should realize that there was something in the child's recognition that Walter's robot lacked.

That is why Zazzo resisted taking the behaviorist imperative as an absolute. As he noted later, his methodological restriction "at the level of observed facts" should not be seen "as respecting a taboo, the definite exclusion of the fact of consciousness."[11] The rise of "consciousness" (*conscience*) implicit in mirror recognition, Zazzo thought, indicated a world apart from simple "sensory-motor intelligence." It was a world of "signification" and "representation."[12] Human infants were set apart from robotic tortoises, because they could recognize images as images.

The need to combine epistemic modesty with the goal of understanding consciousness explains why Zazzo latched onto Gallup's and Amsterdam's mark tests (*l'épreuve de la tache colorée*). It was, he thought, "ingenious in its simplicity," and the "most rigorous method" for determining the existence of mirror recognition.[13] Zazzo would subsequently refer to both Amsterdam and Gallup across his publications.[14]

The discovery of the mark test allowed for a significant shift in Zazzo's work. Having discovered a secure means of determining whether self-recognition had taken place, whether the subject had identified itself in the mirror, Zazzo focused his efforts on disentangling the various elements and thus the meaning of mirror recognition. In a set of experiments conducted over a year with his collaborator Anne-Marie Fontaine, Zazzo filmed thirty twin children between ten and thirty-three months old in front of the mirror and then opposite the twin (mostly identical, though with fraternal twins as a comparison) separated by a glass panel. The researchers could switch between the two situations by sliding the mirror in and out of the glass partition between the two children (fig. 1.1 and fig. 1.2).[15]

Zazzo drew on Amsterdam's complex description of the stages of mirror recognition, agreeing that consciousness did not appear in one go. The "appropriation of the self-image" was an "event that inscribes itself into a long history."[16] Amsterdam, if we remember, had described the transformation from social reactions through avoidance before recognition at twenty-one to twenty-four months. But Zazzo reinterpreted the data in light of the "distress" (*désarroi*) he had noted in his own child when he had recognized his mirror image in 1947. Even in Amsterdam's final stage, where ten out of sixteen children reached to their noses to touch the mark they had first seen in the mirror, a full fifteen still manifested avoidance reactions. This persistence of a purportedly earlier stage, he thought, cast doubt on Amsterdam's "recognition" interpretation. Rather than being the moment when self-recognition emerged complete on the scene, it only offered "the first signs." The problem, Zazzo thought, was a common one. Like many other researchers, Amsterdam had been looking out for the moment of self-recognition and so had brought the experiment to an end after the first behavior that could be plausibly interpreted that way.[17]

For Zazzo, Amsterdam's reading didn't take into account the peculiarities of the mirror. When we see ourselves in the mirror, we confront a self that is very different from the one we normally encounter. The child did not simply look in the mirror and see a familiar face; it took time to take the strange and unfamiliar image and reconcile it with the child's sense of self. This was the process Zazzo caught on film.[18] At first the child was charmed by its mirror image (Darwin's smile), seeing it as another child—Zazzo paid particular attention to the way the child "tapped" on the mirror, as it tapped to its twin through the pane of glass.[19] At the beginning of the second year, however, the child stopped tapping on the mirror, while continuing to tap on the glass pane.[20] At this stage, Zazzo thought, it had begun to recognize the peculiarity of the mirror (here Zazzo referred to Gallup's experiments with Siamese

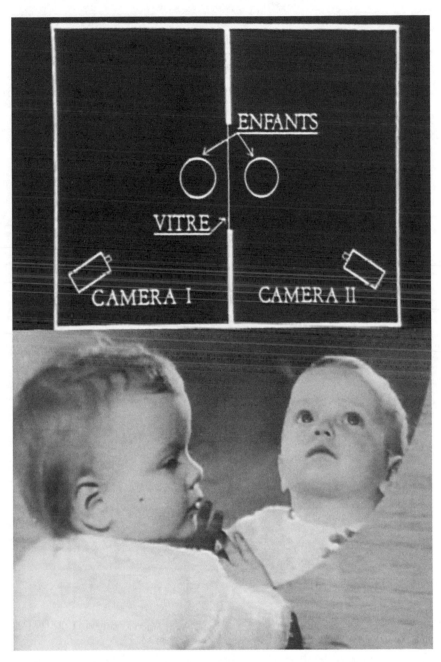

FIGURE I.1. Screenshots from a film featuring René Zazzo's work. The window pane could be replaced with a sliding mirror. René Zazzo, with Anne-Marie Fontaine, *A travers le miroir—étude sur la découverte de l'image de soi chez l'enfant* (1973, 1 janvier), CERIMES, Canal-U, https://www.canal-u.tv/40869, video, 2:06 (*top*), 1:07 (*bottom*).

FIGURE I.2. The girl engages with her twin sister on the other side of the glass pane. Anne-Marie Fontaine, *L'enfant et son image* (Paris: Nathan, 1992), on 33.

fighting fish): "The specular image is not perceived in the same way as an object seen directly or through a glass pane." The peculiarity of the mirror arose by the strange concordance of the child's body and that of the mirror image, an "echo, duplication, circular reaction effect."[21]

The peculiarity of the mirror, its strangeness, explained why the child sought to avoid it. But whereas for Amsterdam this was a stage left behind at the moment of self-recognition, for Zazzo the two overlapped. As Zazzo argued, through playful gestures in front of the mirror, the child slowly came to relate aspects of the mirror image to itself: first its fingers, then hands and other visible parts of the body, and then its face, which proved more difficult given that it was normally visually inaccessible.[22] That was why, at around two years, the child would progress beyond social reactions to the mirror image and pass the mark test. And yet this was an identification with a difference. According to Zazzo, at this stage and despite recognition, the child still believed that its reflection was a real object existing in the world, albeit with a peculiar relationship to its own body. The child continued to apply only its name to its image, even after coming to use the pronoun *me* to describe itself: "The identification is therefore not perfect."[23] That was why for a while after the first signs of recognition, the specular image was still "affected by uncertainty and worry."[24] Complete recognition occurred only when the child went beyond the realist illusion, and recognized the image as an *image*, as a "symbolic expression."[25] The avoidance reaction then dissipated, because the idiosyncrasies of the mirror made sense. The child had constructed "a virtual space where the real space is reflected."[26]

The mark test then allowed Zazzo to shift his attention. Having sidestepped the problem of interpreting mirror behavior, working out if in fact the child had recognized its reflection, Zazzo could now interrogate the nature of that recognition. What did it mean to identify with the mirror image? What were we to make of the differences between the mirror image and the self it reflected, not least the reflection's status as an *image*? Distress, for Zazzo, arose precisely because the child had sensed but not fully understood the peculiar status of the mirror image, that its own body was real and its reflection was virtual. Mirror identification was in fact a form of misidentification.

To grasp the difference of the mirror, Zazzo mobilized a language of signification and the "symbolic" that, while pointing back to the project whose failure had led to the mirror test, gestured to some of its most important later developments. It was also, in France, a clear evocation of a figure who, since the 1960s had become a dominant presence on the Parisian intellectual landscape: Jacques Lacan. The terminological overlap can be traced to Zazzo's reading and work with his mentor at the Laboratoire de psychobiologie de l'enfant, Henri Wallon, who was also, as we will see, a significant influence on Lacan. Wallon in fact had encouraged Zazzo's own interest in psychoanalysis. After completing his doctorate at the Sorbonne in 1933, Zazzo had not initially intended to set off to Yale and work with Gesell. He had wanted, rather, to study with Sigmund Freud in Vienna. Wallon had only scotched the plans as the political situation worsened: "This is not the moment . . . to go to Austria."[27] Nevertheless, Zazzo continued to think of his work as proximate to psychoanalysis; on occasion he described himself as a Freudo-Marxist.[28]

One might have thought that the similarities in approach would have endeared Zazzo to the Lacanians, and certainly several attended the 1973 conference where Zazzo first presented his ideas on mirror self-recognition. But he garnered a hostile reaction, because of a familiar dispute that, as we will see, had significant ramifications. For Zazzo had challenged Lacan on the question of when, exactly, mirror identification was supposed to take place.[29]

PART II

Misidentifications

PLATE 1 (fig. 01). Jan van Eyck, *The Arnolfini Portrait*, 1434. The mirror also has a revealing function: two people are entering the room, one of whom could be the painter. Source: Wikimedia Commons.

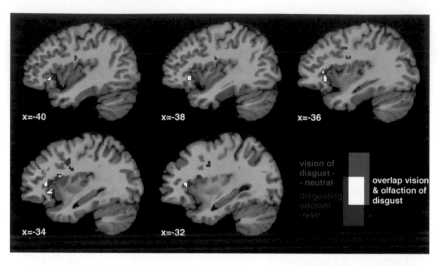

PLATE 2 (fig. 8.6). Disgust and mirror neurons: sagittal sections through the averaged left hemisphere of the fourteen participating subjects. The white patches indicate the overlap of the vision and olfaction of disgust, which are located in the insula. B. Wicker et al., "Both of Us Disgusted in *My* Insula: The Common Neural Basis of Seeing and Feeling Disgust," *Neuron* 40 (2003): 655–64, on 660. Reprinted with permission from Elsevier.

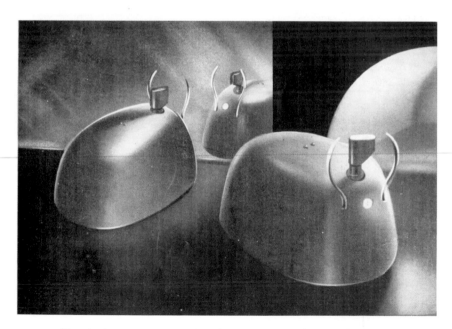

PLATE 3 (fig 3.2). The two tortoises in the exhibition catalogue for the Festival of Britain. The caption reads: "Mechanical 'animals' can be made to steer themselves towards the light." Note that one of the tortoises was presented as "looking" into a mirror. 1951 Exhibition of Science, South Kensington, Festival of Britain, Guide-Catalogue. Dr. W. Grey Walter. Advice and supervision of Model of Electrical Tortoises. With kind permission from the National Archives, ref. WORK25/230/C1/A1/3.

Grasping observation vs object observation

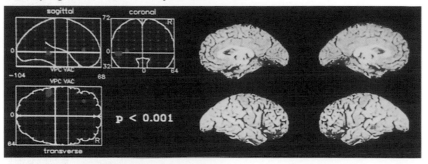

Object prehension vs object observation

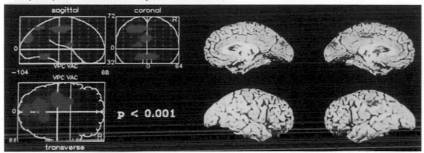

PLATE 4 (fig. 8.8). The inferotemporal cortex and the inferior frontal gyrus were activated by grasping observation (*top*: second row of brains), no activation was found in the inferior frontal gyrus for grasping execution (*bottom*: "object prehension"). G. Rizzolatti et al., "Localization of Grasp Representations in Humans by PET: 1. Observation versus Execution," *Experimental Brain Research* 111 (1996): 246–52, on 248. Reprinted by permission from Springer Nature.

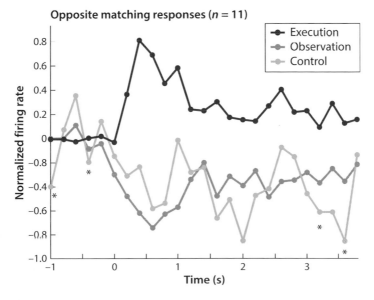

PLATE 5 (fig. 8.9). Super mirror neurons showing inverse mirroring. Reprinted from Roy Mukamel et al., "Single-Neuron Responses in Humans during Execution and Observation of Actions," *Current Biology* 20 (2010): 750–56, on 754. With kind permission from Elsevier.

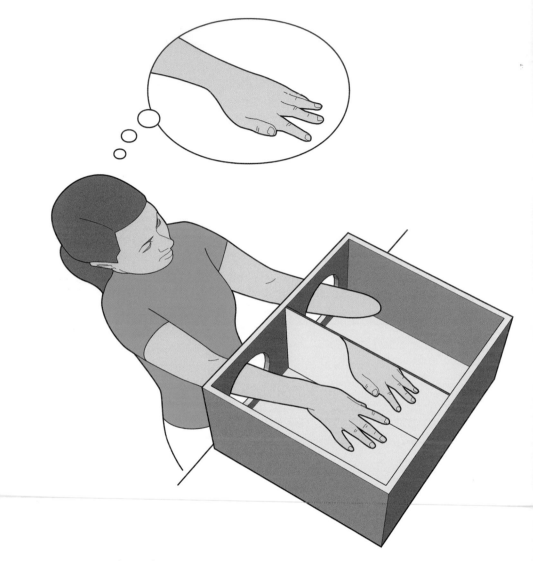

PLATE 6 (fig. 8.10). V. S. Ramachandran's mirror box. V. S. Ramachandran and Diane Rogers-Ramachandran, "It's All Done with Mirrors," *Scientific American Mind* 18, no. 4 (August/September 2007): 16–18, on 17.

5

The Mirror Test That
Never Happened

LACAN, THE EGO, AND THE SYMBOLIC

THE CHILD IS CLEARLY YOUNG. Unable to walk, it is held upright in a walker, or perhaps by an adult, directly in front of the mirror. It moves its hand, noting how the reflected image tracks its movements precisely. This quickly turns into a game, and the child uses it as an opportunity for experimentation, testing the relationship between the mirror image and the real world, both its own body and the other people and things it sees reflected there. Then comes a moment of great joy. The child leans forward in the walker and gazes at the whole image in one go, as if trying to fix it in its mind. It appears drawn to something in the mirror by a powerful psychic force.

This scene follows the description given by the man most closely associated with the mirror self-recognition test in modern thought: the French psychoanalyst Jacques Lacan. In his "mirror stage" paper from 1949, Lacan presented the moment of specular recognition as a key step in the child's mental development, marking the emergence of the ego, and forming the basis of what Lacan called the "imaginary order." The child, who experienced its body as an uncoordinated chaos, unable to control its movement fully, looked in the mirror and saw it as a coherent whole. Fulfilling the child's need for wholeness, Lacan argued, the image becomes an object of desire, hence the "flutter of jubilant activity" upon recognition.

The opposition between the unity of the mirror image and the fragmentation of the real body made the mirror self fictitious, but it did not make it any less potent. In the mirror the child first came to consider itself a self, an individual with particular characteristics.[1] Nevertheless, as a fiction, the mirror image and the ideal self it engendered could never be fully aligned with reality; it was profoundly alienating. As Lacan argued, the mirror image "situates the agency known as the ego, prior to its social determination, in a fictional direction that will forever remain irreducible for any single individual or, rather, that will only asymptotically approach the subject's becoming, no matter how success-ful the dialectical synthesis by which he must resolve, as *I*, his discordance with his own reality."[2] Consequently, Lacan posed psychoanalytic work, both his own and that of his growing school, as an attempt to navigate and even dissolve the fixations of the ego and to let the unconscious speak for itself. Lacan de-veloped these ideas about the mirror stage in the 1930s, and published them in the late 1940s. But by the mid-1950s they were folded into his broader re-reading of Freud, which emphasized the linguistic aspects of psychoanalysis and was governed by the idea that the unconscious was structured like a lan-guage. According to this view, the "imaginary" order of the mirror stage was one part of Lacan's famous "hérésie" (RSI), along with the "real" and the "sym-bolic" orders.

Read within the context presented in this book, Lacan's account of the mir-ror encounter is both familiar and strange.[3] The hand games recall those de-scribed by Gordon Gallup and Ann Bigelow. The moment of joy is reminiscent of Doddy's smile. But Lacan's presentation also has a number of oddities. First, over the course of the twentieth century, as we have seen, the nature of the mirror test had encouraged an epistemic modesty on the part of most practi-tioners. Even with the mark test—which wouldn't be formulated by Amster-dam and Gallup for another twenty years—it was frankly quite hard to determine precisely what was occurring in the mirror encounter, especially since it gener-ally involved, for Lacan as for his predecessors, nonlinguistic creatures: the infant and the animal. Lacan, however, lambasted researchers for their "timid-ity" in refusing to draw full conclusions from their experiments, and settling for the bare assertion of facts detached from broader theories.[4] In contrast, he felt confident enough to insert his account into a vast and subtle theoretical system, and he drew on the mirror test to make some profoundly important but also quite extravagant claims about the human self.

The second oddity might appear minor, but it has far-reaching implications. As René Zazzo argued, Lacan got basic facts wrong, not least that the moment of recognition could occur as early as six months. Though there was consider-able variation, as we have seen, most major contributors to the tradition in the twentieth century had located the moment at which the child identified with the mirror image around the child's second birthday.[5] Lacan's dating cannot simply be put down as a slip. For, it is only by shifting the moment of mirror identification that Lacan is able to argue that it occurs at a moment of significant motor incoordination, and thus make a convincing case that the unity of the mirror image is fictitious.[6]

The two oddities allow us to hazard a surprising hypothesis: Because Lacan ignored many of the practical difficulties of mirror experimentation and erred on some of its most obvious details, one must question whether he had actually ever studied a child's reaction directly. The most famous mirror test in history might never have taken place.[7]

Wallon and the Mirror

Lacan embedded his work into the mirror tradition predominantly through his (generally tacit) engagement with the French psychologist Henri Wallon, whom we have already encountered as Zazzo's mentor. Although almost entirely unknown in the anglophone world, Wallon was one of France's most prominent psychologists in the first half of the twentieth century. He had obtained an elite education at the Lycée Louis-le-Grand and the École normale supérieure, studying philosophy and psychology, and passing the agrégation in 1902. He then turned to medicine, working at the Salpêtrière Hospital with, among others, the histologist Jean Nageotte, whom he cited as an important influence. He held various positions at pediatric psychological clinics, before being named Pierre Janet's successor at the Collège de France in 1937, the highest accolade a psychologist in France could receive.

In the interwar period, Wallon dedicated significant time and energy to the mirror test in his writings, and he grappled extensively with some of the figures I discussed in chapters 1 and 2: Wallon read Darwin, Preyer, and Charlotte Bühler, among others. Indeed, his most important publication, *Les origines du caractère chez l'enfant* (1934), is packed with references to key individuals from the earlier movement with close readings of many of their mirror experiments. Wallon did not engage with the literature because he lacked the time or inclination to conduct the experiments himself; though he did not have children, a few scattered references throughout the book show that he was not averse to testing his hypotheses directly. Nor did he do so in order to ratify the conclusions of the tradition. Instead, he sought evidence for his own recasting of the mirror test in the very material that had authorized their interpretations. In particular, he suggested that while they might have seen a dawning recognition of the self in the behavior of children they had observed, for him that behavior heralded a different but no less consequential development. According to Wallon, the moment children came to grasp the significance of the mirror image, they entered an ethereal world of disembodied ideas. The mirror was a gateway into what he called the "symbolic."

Like many of the figures we have encountered thus far, Wallon's work was driven by an acute sense of the dangers of introspection. He worried that, as conscious adults, we all too easily project our own responses to the mirror onto the behavior of prelinguistic children and animals. Such an approach blinded us to the foreignness of our cognitive beginnings and imposed a linear teleology, where children and animals were taken to manifest more primitive versions of our adult human faculties. To avoid such linear readings, Wallon demanded that we examine modes of consciousness by paying attention solely to their "actual manifestations, that is, the concrete way in which they manifest themselves."[8]

In rejecting this linear form of teleology, Wallon did not disavow all forms of development. As he remarked at the beginning of the book, despite the vastness of child psychology, "to want to confine oneself too tightly there would only be to end up making inventories, chronological enumerations, simple descriptions."[9] For Wallon, facts of child psychology gained meaning only in so far as they were understood as part of a broader whole defined by maturity: "The species can only find its purpose in the adult type . . . the infant tends toward the adult like the system toward its equilibrium."[10] In addition to the adult telos, Wallon added a social one: "To split man from society, to oppose, as is frequent, the individual to society, is to decorticate the brain."[11]

As this last formulation suggests, Wallon followed earlier figures like Preyer in deploying a neurological hermeneutic; the physiological development of the brain provided the means of recognizing the alterity of children, while relating their experiences to that of adult humans. Wallon noted that, in his early work, he had tried to follow "the progressive maturation of nervous centers" at the same time as the "successive predominance of different behaviors."[12] For Wallon, human development could be understood as the slow association of sensory and motor systems, knitted together by an increasingly complex nervous system. Whereas at first the child experienced only a chaotic set of sensations and lacked any form of motor coordination, over time as consistent correlation became hardwired in nervous connections, experience and action in the world allowed the child to experience it as a cohesive and ordered whole. That is why Wallon was particularly interested in the processes of myelinization, which provided the physiological infrastructure of association between the various sensory fields, and he discussed the formation of the pyramidal tracts (*faisceau pyramidal*), which transmitted motor commands to peripheral neurons.[13]

The first significant associations occurred in the third month, when the child began to be able to coordinate the movement of head and eyes. Around the seventh month, Wallon asserted, the child managed to gain more consistent control of its body (though he admitted that the range was very large, and that complete control of the body occurred much later).[14] At that time, as he wrote, "the agitation ceases to be completely diffuse and the different parts of the body cease to intervene randomly or through a simple motor impulsivity."[15] By about one year, this sense of coherence began to expand beyond the body, and the child started to make sense of the objects in the outside world, entering what the gestalt psychologist Karl Bühler had called the "chimpanzee stage," at which it began to use objects as tools.[16]

The child's growing awareness of its own body, produced through the progressive association of neurons, set the stage for Wallon's treatment of the mirror test. While consideration of the nervous system knitting itself together led previous figures including Preyer to look into the mirror for evidence of a self-concept, it encouraged Wallon to ask how the subject separated itself from its environment. It is a subtle distinction, but one with important consequences. Take Wallon's discussion of animal responses to the mirror.[17] Preyer's duck could not have found consolation in its mirror image after the loss of its companion, Wallon argued, because ducks did not have a sense of the individual. Rather it experienced other ducks according to a "syncretic intuition." As he wrote, "it is a state of sensibility anterior to that where the person knows how to dissociate herself from the environment and distinguish, between her impressions, those that refer to herself, and those that refer to the external world."[18] In short the very question of whether the animal saw itself or another in the mirror image presupposed a distinction that the duck was unable to form. The duck experienced its mate's absence not as the lack of presence of a companion but rather as akin to "an amputation."[19] Consequently, the mirror image did not come to replace the old mate; it reinstituted the sense of wholeness to which that mate had previously contributed. So too, in his own experiments on fox terriers, Wallon argued that they did not distinguish their own feelings from that of the mirror image, which was more like an "echo, without any feeling of doubling."[20] Some animals did not fall into this trap. Wallon remarked that "higher apes" showed a "reaction . . . at a much higher level."[21] Upon encountering a mirror "they instantly put their hand behind [the mirror], and showed their anger at having found nothing and refused to look further. It seems like a veritable act of recognition." But this disappointment merely led them to ignore the mirror in the future.[22]

Humans were different. For them the first milestone came at around six months. This was when children started to "turn around to the person whose image they perceive in the glass."[23] This reaction couldn't be understood as a simple connection between the appearance of the image and the presence of another, linked together by habitual association. Rather for Wallon it was "the verification of a relation, an act of recognition."[24] That is, the child had begun to develop a rudimentary understanding of how a mirror worked and had begun to organize in experience the relationship between various parts of its visual field. Similarly, at around eight months (and here Wallon cited Darwin), the child started to turn to the mirror when its own name was called, in this sense recognizing itself.[25]

In comparison to the rest of the tradition, Wallon's dating of self-recognition is remarkably early.[26] But because he was interested in the process by which the child distinguished its own body from the rest of the world, and not in the self-concept, he did not consider this to be a decisive step. For Wallon, the trouble with the mirror image was that it was spatially separated from the body; though children were "capable of perceiving a relationship of resemblance and concomitance," at that moment "they did not yet grasp their true relations of subordination."[27] In particular, and crucially for Wallon, the child still attributed reality to the mirror image and gave priority to the real person in mirror experiments (e.g., by turning to them), only because sounds emanated from that person rather than from the image.[28] That is why weeks after recognizing that the reflection of another person indicated that person's presence, the child still tried to grasp its own mirror image, as if it too were a physical object in space.[29] At this point, the child subscribed to the "animist" idea that it inhabited two bodies at once: the body in the real world that it could feel, and the body in the mirror that it could see. The child was simultaneously "in the space that is one with its proprioceptive impressions and [in the space] animated by her exteroceptive image."[30] That is, mirror recognition reproduced an earlier error of thinking exterior objects were part of the child's body.

To mark the borders of the body correctly, Wallon thought, a child needed to be able to associate its proprioceptive with particular exteroceptive sensations (essentially its bodily feeling with a visual experience of its body) in a way that would allow it to distinguish self and other spatially. For this, mirror recognition did provide a valuable contribution. Without a reflecting surface, Wallon argued, it was impossible for the child to gain "a homogenous and coherent image of the whole body." The child could see its body, but only in "fragments and never assembled": first its headless torso, then one shoulder, followed by another, after which it could bring a foot into its visual field. The result was a "fragmented vision [*vision parcellaire*]."[31] In the mirror, in contrast, the child saw its entire body in one go and in a complete outline, which cut off the body as a whole from the rest of the world.

While the infant was still happy with an animist doubling, however, it was not able to exploit this experience to mark the limits of the self. The unified exteroceptive self was separated spatially and was distinct from the proprioceptive self. To bring the two together, the child had to be able to (as Wallon suggested later in the book) treat the exteroceptively experienced mirror image as "clothing" [*vêtement*], remove it mentally from the mirror world and drape it over the proprioceptive experience of the body. Only thus could the unified image furnish a clear boundary between the bodily interior and the outside world. To perform that operation, the child needed to see the image as an image, subordinate to the body it reflected and without bodily reality itself. That is, the child had both to "accept images that only had the appearance of reality" (the mirror image) and to "affirm the reality of images that elude perception" (the direct visual unity of the body), in order that it could see the former as a reflection of (and thus visual access to) the latter.[32] Only displaced in this way could the mirror image play a role in the broader process of organizing sensory experience. This, according to Wallon, started to occur at around one year, when the French psychologist Paul Guillaume's daughter reached up to her hat after seeing it in the mirror (an early version of the mark test).[33] The crucial point here is that the girl had "emptied" the mirror image of existence, and thus was able to transfer it onto "her proprioceptive and tactile self [*moi*]; [the image] is nothing but a system of references, apt to orient gestures toward the particularities of [the child's] own body that it indicates."[34] When, at sixty weeks, Preyer's daughter turned back to see her mother, this for Wallon was qualitatively different from the similar gesture eight months beforehand (the first milestone). Now the child had "come to recognize clearly the irreality and the purely symbolic character" of the mirror image.[35] For Wallon this development was fully secure only by the time the child reached its third birthday.[36]

The new sense of its own body, for Wallon, was an important psychological achievement, which formed the basis for later developments; it would "integrate itself more or less tightly into the notion that will develop in the child the awareness of its moral personality toward other people."[37] Consequently, the consciousness of the body came to be shaped by social and civilizational factors. This was the topic of the final section of the book, focused not on the "consciousness of one's own body" (*conscience du corps propre*) but rather the "consciousness of the self" (*conscience de soi*). The narrative followed a similar arc to the one concerning bodily experience, moving from an indistinct chaos, through a moment of identification, to juxtaposition and ultimately symbolization. Wallon described an early syncretic sociability, where the other was seen as an extension of the self. It was only in the sixth month that within this syncretism the child started to make a sense of itself and rivalry with others of the same age. This rivalry was, however, built on a form of identification, and here Wallon drew on Charlotte Bühler, who suggested that a rivalry could develop between two children as long as they did not differ in age by more than two and a half months.[38] For instance, in discussing jealousy—something, he suggested, that arose only at around six months—Wallon wrote that it would be a mistake to suggest that the child had an idea of "personalities clearly distinct from its own." Rather the moment demonstrated "still participation, but participation of a contrasting nature that announces the moment of individualization. Torn between two poles whose rapprochement within its own sensibility helps resolve the opposition, the subject is quite close to feeling the necessity of focusing on one, and crystallizing a person different from it around the other."[39] This occurred fully only at the end of the third year, at which point, Wallon insisted, the child developed a sense of self as absolutely distinct in a world of others, built on its solidified sense of the distinctiveness of its own body.

As we can see, crucial for the development of the self was not the recognition that the mirror image was a reflection of the child's own body, but rather the recognition of the image's curious ontological status, and thus the discovery "of purely virtual systems of representations."[40] That is why Wallon presented this moment as a "prelude to symbolic activity, through which the mind comes to transform sense data into a world."[41] The sheer importance of representation in Wallon's account stages the return of a set of claims that had been absent from the mirror tradition for many decades. As we have seen, in the latter half of the nineteenth century, Darwin's baby-diary work had been taken as evidence that the language demarcator was no longer operative, and the mirror test became an uneasy substitute. In contrast, for Wallon the mirror stage was primarily about the question of representation and by extension language. As Wallon wrote, the development of language required "representational thought," which as "delimited by means of signs, gives rise to an opposition between the same and the other, the like and the unlike, the one and the many, the permanent and the transitory, the identical and the changing, the stationary and the moving, and the being and the becoming."[42] According to Wallon, the development of language, higher-order thought, and the development of a sense of self were thus deeply integrated. Wallon went further:

> Language is, in fact, the essential step biological evolution has enabled man to make. Within his nervous system are centres that make speech possible. Once he can give names to things, and to the relations between things, he is able to evoke them in their absence, to combine at will the images he has of them, to transmit his knowledge, and to receive the knowledge of others. From this arises the possibility of civilisations increasing their heritage from one age to the next.[43]

Wallon also cited the example of a monkey who showed itself at a similar stage of development and mental abilities as a child, but was quickly left behind once the child began to develop language.[44]

The thing that set humans apart for Wallon was not self-recognition, at least not directly. Wallon was not particularly interested in the question whether the subject identified the image in the mirror as itself or as another individual. For him, what was exciting and interesting about the mirror test was that it wasn't about identification at all. The most salient characteristic of the mirror image was that unlike the self it reflected, it had no physical existence. We understood the mirror, according to Wallon, not when we recognized ourselves in the image, but rather when we recognized our mirror selves as unreal. Unlike the previous tradition that had been focused on the question of identification, whether and how the child recognized the image as itself, Wallon brought to the fore *misidentification*, prizing apart image and self.

Lacan's Mirror

Lacan's paper "The Mirror Stage as Formative of the *I* Function as Revealed in Psychoanalytic Experience" was published in the 1949 volume of the *Revue française de psychanalyse*; but its central ideas went back to before the war, when Lacan presented at the fourteenth congress of the International Psychoanalytical Association (IPA) held at Marienbad (today Mariánské Lázně), a Bohemian spa town, in August 1936.[45] Lacan was at the time a young psychiatrist working at the Sainte-Anne hospital in Paris. Despite his youth, he was already recognized as an important thinker in France. In literary and artistic circles of Paris, his 1932 thesis, *De la psychose paranoïaque dans ses rapports avec la personnalité*, was considered an inspiration for its literary style.[46] The book in fact reads like a novel, based as it was on the case of Aimée (real name Marguerite Anzieu, née Pantaine), who in April 1931 attacked the famous Parisian actress Huguette Duflos with a knife, was then hospitalized, and became Lacan's patient. In Lacan's reading, when attacking Duflos, Marguerite Anzieu struck her own ideal self, for she aspired to be a famous writer and public figure. Since the attack was ultimately aimed at herself, it brought her no relief. She gained satisfaction and thus respite only upon being tried and punished.[47] As Lacan wrote, "The nature of her cure demonstrates, we believe, the nature of her illness."[48]

Within psychiatry, Lacan was burdened with the hopes and expectations attached to a new intellectual leader. But he was still a relative newcomer in the world of psychoanalysis. In 1936 he was in the middle of his training analysis with Rudolph Loewenstein and was on the verge of entering the titular ranks of the Société psychanalytique de Paris (SPP).[49] Thanks to his marriage to the well-heeled Marie-Louise Blondin, he was able to open his own private practice, while continuing to see patients at Sainte-Anne. Thus when Lacan traveled to present his paper at the Marienbad congress, he was a young and unestablished psychoanalyst, and his radically provocative remarks were not well received. A version of the paper given two months earlier to the SPP seems to have aroused a confused and rather negative reaction, and in Marienbad the talk was cut short by the chairman, Ernest Jones (though Lacan reported that the Viennese group was more open to his ideas).[50] In response to this dismissive treatment, Lacan stormed out of the congress, deciding to spend his time instead taking in the Nazi spectacle of the Berlin Olympics.[51] He never submitted his paper to the conference proceedings. As a result, the paper was listed in the *International Journal of Psycho-analysis* among the proceedings but was not included, not even in summary.[52]

Nonetheless, when Wallon, a man who showed interest in psychoanalysis but was by no means part of the psychoanalytic community, was asked by the historian Lucien Febvre to edit the psychology volume for the new *Encyclopédie française*, Lacan's liminal status, as a psychiatrist—Wallon had met him several times between 1928 and 1934 at the Société de psychiatrie—with a newfound interest in Freud, made him an obvious contributor.[53] His article on family complexes was one of several dealing with psychoanalysis. Others were written by Edouard Pichon, founding member of the SPP (in 1926) and proponent of the "chauvinist" strand,[54] and Daniel Lagache, who was charged with writing on sexuality.

Though Wallon is not cited at any point, Lacan's paper drew on many of the arguments Wallon had developed, to such an extent that the Lacan biographer Elisabeth Roudinesco comes close to charging Lacan with plagiarism.[55] Just as Wallon worried about the projection of ideas developed through introspection onto the prelinguistic child, so too did Lacan worry about projecting structures of a an adult mind, in particular the Oedipus complex, onto a child not yet developmentally mature.[56] This notion of development was also constantly undergirded by reference to neurology, and like Wallon he referred to the delayed myelinization of nervous centers, which justified Lacan's structuring argument about human's "specific prematurity at birth."[57] We also see a similar trajectory and chronology, where the child's early development set the groundwork for its relationship with others; for both Wallon and Lacan, the way the child developed before one and a half years old rendered it susceptible to a range of different social and familial influences afterward.

Lacan cleaved closest to his older colleague on the question of sociability. His discussion of sibling jealousy drew heavily on the themes and questions of Wallon's book, including Charlotte Bühler's argument that jealousy occurred only when the two children are separated by less than two and a half months.[58] Lacan sided with Wallon against the majority of other psychologists, arguing the central feature of jealousy was not competition or antagonism but identification: "Each partner confounds the domain of the other with its own and identifies with him."[59] In the early stages the child identified with the mother and what was beyond itself, what Wallon had labelled "syncretic sociability," and it was only later that the difference between the two was noted. Consequently, the infant's world "does not contain the other."[60] Then Lacan discussed the emergence of what he would later call the "social I," when the child competed with the sibling for a toy, and thus folded the imaginary ego into a social setting, something that was completed, as for Wallon, only by the age of three.[61]

The section on the mirror seems, at least at first glance, also to be consonant with Wallon's ideas. We have a similar argument about the comparative advantage, but developmental retardation, of higher apes with respect to the child at around one year old, even making the same reference to Karl Bühler's "chimpanzee stage." And correlatively, we have the absolute centrality of the mirror stage for demarcating the distinctively human.[62] Most importantly, Lacan made claims about the productivity and the fictionality of the mirror image that had been central to Wallon's account. Like Wallon, he placed emphasis on the fact that the mirror image was *not* the child. As we have seen, according to Wallon, at around eight months the child identified with the image, erroneously, not yet recognizing its irreality. Lacan concurred: this was a time when the child "confounds itself with [the mirror] image."[63] This process, for both men, then laid the groundwork for the emergence of a sense of self.

These similarities, however, should not blind us to two significant shifts in Lacan's analyses. First, for Wallon the identification with the mirror was not in the first instance about consciousness of the self, which Wallon located at about three years (corresponding approximately to the emergence of Lacan's "social I"), but rather about "consciousness of the body." Lacan did not make definitive claims at this stage, but he was clear that the unity of the mirror image, intruding into the child's psyche, "will contribute . . . to the formation of the ego [*moi*]."[64] Second, Lacan altered the timing. While Lacan did accept Wallon's dating of eight months in a relatively obscure 1953 paper,[65] he shifted it two months earlier in both the 1938 encyclopedia entry and the 1949 "Mirror Stage" article (as well as its reedition in the 1966 *Écrits*).[66] Lacan's description of this moment, which produced a "jubilatory outburst of energy," might be a reference to Darwin's description of Doddy's first response to the mirror (a smile at four and a half months), which Wallon himself discussed. But even given the disparities in the dating and the drama of the description, it is not clear that this reference would serve the purpose Lacan wanted. As we saw, Darwin made no mention of mirror identification at that point. The shift in the dating of mirror identification is matched in Lacan's treatment of jealousy, which Lacan used in 1938 as a frame for his analysis of the mirror stage.[67] Whereas Charlotte Bühler and Wallon both noted the emergence of jealousy at the ninth month, Lacan placed that at six months too.[68]

To understand why these small changes in the chronology might be significant, it is worth quoting the relevant section on the mirror test from the 1938 text at length. The reaction to the mirror, Lacan asserts,

is nothing but the consequence of the prolonged incoordination of apparatuses. From it results a state constituted mentally and affectively on the basis of a proprioception that treats the body as fragmented [*morcelé*]: on the one hand, the interest of the psyche finds itself displaced onto tendencies that aim at some reconstitution [*recollement*] of the body itself; on the other hand, reality, subordinated initially to a perceptive fragmentation, whose chaos reaches up to the categories, "spaces," for example, as disparate as the successive states of the child, organizes itself in reflecting the forms of the body, which provides in a certain way the model for all objects.[69]

Superficially, the discussion again seems to parallel Wallon's. Where Wallon talked about a "fragmented vision" (*vision parcellaire*), Lacan discusses a "perceptive fragmentation" (*morcellement perceptive*), both of which informed categories of space and could be overcome through the process of looking into the mirror.

A closer look reveals a number of significant differences. As we saw, for Wallon, the mirror image was able to counteract the disunity of the child's *exteroceptive* experience of its body, the impossibility of the child seeing all parts of its body in one go. This is important because, for Wallon, the mirror image was crucial in helping distinguish self from other, and the outlines of the mirror image provided the necessary schema to do this. By "clothing" the (proprioceptive) experience of the body in the mirror image, the child was able to separate it from the nonbodily.[70] But Lacan was not interested in the way the mirror image sharpened the lines between self and other. Rather, he wanted to analyze the way it blinded the subject to divisions within the self. That was why he placed his greatest emphasis not on exteroceptive fragmentation, but rather "a *proprioception* that gives the body as fragmented."[71] It is true that Wallon also discussed proprioceptive incoordination in his book, but that was in an earlier chapter, describing an earlier stage of child development. According to Wallon, by the seventh month—crucially before he thought the child was able to identify with the mirror image—the child was discovering "the unity of its body, of which the various parts are capable of voluntarily executing the same act."[72] Lacan, in contrast, extended this incoordination into the second year: "The study of behavior of early childhood allows us to affirm that extero-, proprio-, and interoceptive sensations are not yet, after the twelfth month, sufficiently coordinated that consciousness of the body can be achieved, nor correlatively the notion of what is exterior to it."[73]

This liberal use of the mirror tradition, the mixing up of the various stages, to shift mirror-image identification back to a time where the child was still largely unable to control its body, continued in later versions of the mirror-stage paper after the war. Lacan presented a revised version to the IPA at its sixteenth congress in Zurich, in July 1949, which was then published and reprinted as the version we know today. The 1949 version was published at the height of the excitement for existentialism in France, and Lacan used it to articulate his relationship to Sartre and Simone de Beauvoir.[74] But the paper also reaffirms

Lacan's indebtedness and challenge to Wallon. Lacan made Wallon's arguments about motor coordination and the pyramidal tracts even more prominent (though again not naming the older man).[75] He rhetorically linked these to the mirror stage, presenting it as a moment of significant motor incoordination, by introducing an element that was absent in the 1938 version: the dramatization of the child's being held upright in front of the mirror by a "prop, human or artificial (what, in France, we call a *trotte-bébé* [a sort of a walker])," with which I opened this chapter.[76]

A two-month disparity in the dating of a development that manifested considerable variability from child to child might not seem much.[77] But we must remember that we have been comparing Lacan's account to Wallon's discussion of mirror-image *identification*, which Wallon had placed considerably earlier in child development than most other scholars. Identification, however, was not sufficient for the mirror image to function as a unifying schema. As Wallon made clear, the process by which the child related the mirror image to its own body—whether to lay out the limits of the self (Wallon) or to lend fictive unity to an uncoordinated motor apparatus (Lacan)—also required symbolization, for which the first steps take place only after the first birthday, and properly speaking even later; this was considerably after the beginning of Lacan's mirror stage.[78] This condition holds even if we consider that for Lacan the mirror stage need not be prompted by the confrontation with an actual reflecting surface, but through imaginary identification with any "semblable." Lacan seemed to recognize the necessity of symbolization. It is significant that in the 1938 article he described the mirror stage under the rubric of the "intrusion complex" (*complexe d'intrusion*); the unity of the mirror image could "intrude" on the mind only insofar as it could extract itself symbolically from the mirror realm, such that the child can recognize it as an image, and thus use it to inform its proprioceptive experience.[79]

The innovation of Lacan's account thus boiled down to two parts: He folded together three separate sections of Wallon's book—the discussion of bodily coordination (mostly in the first seven months) and the two moments of mirror recognition, both identification with the image (ca. eight months) and derealization of the mirror image to construct a schema (twelve to eighteen months). And he fiddled with the dating to construct a presocial moment of symbolic identification caught in the dialectic of jealousy, which occurred at a time of complete motor incoordination. Then he associated them all with a prior moment, described by others like Darwin, of a positive emotional response to the mirror image occurring at six months or even earlier.[80] All this was posited as the process that produced not simply consciousness of the body but the ego itself, something that for Wallon occurred only at three years. The strategy recalls Preyer's recommendations for writing baby diaries: the observations had to be noted down in such a way that they could be cut out and rearranged. Here, however, Lacan was not, like Preyer, rearranging the order to separate out different developmental trends; he was changing the very chronology of the mirror encounter. Given how he played with the chronology, it is very unlikely that he could ever have seen a child identify with the mirror image in the way he describes.

Lacan picked up Wallon's innovative contribution to the mirror-test tradition, a shift from an analysis of whether and how the child identified with the mirror image to an analysis of the pathologies of that identification; but the shift in timing gives these pathologies a different gloss. According to the revised timeline, the child looked into a mirror and confronted an image whose wholeness contrasted not just with its exteroceptive but also with its proprioceptive experience, the "prolonged incoordination of apparatuses."[81] Rather than an overcoming of the visual (but not actual) fragmentation of the body in the mirror image as part of a broader process of distinguishing self and world, the unity of the image became a "fiction" that contrasted with the proprioceptive (and in this sense real) fragmentation of the body. Only with this inversion of the developmental processes could Lacan institute the crucial temporality of the mirror stage, whereby the "subject *anticipates* the maturation of his power" in the image.

A large motivation for Lacan's transformation of Wallon's mirror stage stemmed from differing understandings of the problem of immaturity, and thus what was wrong with mirror identification. Wallon sought to show how the child slowly came to impose order on a chaos of sensory impressions; for Lacan the child was driven to overcome a lack.[82] Lacan's encyclopedia article was framed by this very issue. The section on the "intrusion complex," which contained the passages on the mirror, immediately followed a section on the "weaning complex" (*complexe de sevrage*), which highlighted the sense of loss caused by separation from the mother (something not discussed at length by Wallon). In short, the trauma of weaning, which itself repeated the initial trauma of birth, created a strong desire to recover lost wholeness, and it was this desire that led the child to break from the prison of instinct.[83] This sense of lack was also at the heart of Lacan's reading of jealousy. The identification with the sibling arose not, as for Wallon, from an inability to distinguish self and other, but rather from a sense that something was missing, which could also be traced to the process of weaning and which presented identification as a process rather than a state.[84] We can see why Lacan was so keen to transform Wallon's argument about the child's inadequate view of the body into an argument about the inadequacy of the body itself, and thus why he was tempted to push the moment of identification back in time. This shift allowed him to present the child's "capture" by the mirror image as a process driven by desire.[85]

The recasting of the mirror encounter also bears the mark of Lacan's psychoanalytic training. With important provisos to which I will move shortly, Lacan read into the mirror test a number of ideas that he had already come to embrace in his engagement with Freud. When discussing the mirror stage, Lacan turned again and again to Freud's treatment of narcissism.[86] In his essay "The 'Uncanny'" (1919), Freud had treated narcissism in relationship to the double, which was among the most favored motifs of literary Romanticism— E.T.A. Hoffmann, Adalbert von Chamisso, Oscar Wilde, and Edgar Allan Poe all drew in some way or another on the doubling of the self, often as a mirror image or a shadow.[87] As Freud noted, in these stories, the "Ich-Verdopplung" (doubling of the self) of the double began as "insurance against the destruction of the ego," an "'energetic denial of the power of death'" in the words of Otto Rank, Freud's disciple and author of *Der Doppelgänger* (1914).[88] Freud related the creation of the double to the "primary narcissism" of the child, whose identifications added content to the ego.[89]

This, however, was not the end of the story. Freud noted that after the passing of primary narcissism, the double became the "uncanny harbinger of death."[90] In particular, the doppelgänger could be turned into a kind of censor, a moment of self-critique, which would "treat the rest of the ego like an object":

> The idea of the "double" does not necessarily disappear with the passing of primary narcissism, for it can receive fresh meaning from the later stages of the ego's development. A special agency is slowly formed there, which is able to stand over against the rest of the ego, which has the function of observing and criticizing the self and of exercising a censorship within the mind, and which we become aware of as our "conscience."[91]

In his essay "On Narcissism," from 1914, Freud had explicitly linked this idea of censorship or repression to a secondary form of narcissism, investing in a part of the ego that was subsequently set up as an "ideal." As Freud wrote, this "ideal ego [*Idealich*] is now the target of the self-love which was enjoyed in childhood by the actual ego. . . . What he projects before him as his ideal is the substitute for the lost narcissism of his childhood in which he was his own ideal."[92] We can see here the resonances with Lacan's reading of the mirror stage: the deficiencies of the real being are made up by the ideal double, from which the subject was ultimately alienated.

It would be wrong to see this as the simple translation of Lacan's psychoanalytic commitments onto the mirror test, for the translation had important implications for that theory.[93] For Freud, the alienation of the ideal ego was a product of secondary narcissism *after* the creation of the ego. He would later see this as the basis of the superego. For Lacan, the ego was itself ideal and alienating. In the mirror stage paper, he referred to the "semantic latencies" of narcissism: a young man trapped by the reflected image of his beauty.[94] This translation of the characteristics of secondary narcissism to primary narcissism can be traced to the fact that the mirror test, despite Wallon's attempt to recast its meaning, had long been associated with the rise of self-consciousness.[95] That was why, in contradistinction to Freud, Lacan distinguished the *Ideal-Ich* as object of identification (for instance in the mirror image) from the *Ich-Ideal* as element of the superego.[96] Second, while primary narcissism for Freud was the process that produced identifications with the ego, for Lacan—along the lines of Wallon's reading of the mirror test—narcissism created the ego in the first place. The appeal to the mirror encouraged Lacan to displace Freud's argument about the formation of the superego onto the ego with significant implications for psychoanalysis.[97]

Lacan and the Symbolic Function

My argument that Lacan reoriented the mirror test away from the question of language or the symbolic (the role it had played for Wallon) and back to the emergence and recognition of the self might seem ironic given Lacan's accepted status in the intellectual history of the twentieth century. Lacan is often seen as part of a broader antihumanism that sought to decenter the subject, and his major contribution to psychoanalysis was his "linguistic" reading of Freud. That reading found its canonical formulation in the 1953 Rome Report, where language was introduced as one of the three orders—the real, the symbolic, and the imaginary—that became central to his thought.[98] Of the three orders, the symbolic was the most important because it allowed Lacan to characterize the nature of the unconscious—it was "structured like a language."[99]

Lacan's insight came from the centrality of speech in psychoanalysis, named the "talking cure" by Bertha Pappenheim (case pseudonym Anna O.), along with the symbolic nature of the material Freud dealt with, not least dreams and slips of the tongue.[100] Lacan argued that pathological symptoms should be seen as the product of symbolic displacement, a form of metonymy or metaphor, where one signifier was replaced by another.[101] The symbolic was also intersubjective, in the sense that it preceded the individual (like language) and necessarily involved the other for whom the language was intelligible. That was why Lacan labelled the unconscious the "discourse of the other."[102] This meant, most importantly, that the logic of the symbolic was utterly distinct from biology and the force of instinct. At this stage, it was clear that the symbolic marked humanity's decisive step over animals, who had a simple relation to the imago—thus returning fully to Wallon.[103]

At this time Lacan argued for the imbrication of the imaginary and the symbolic. The latter was the system that held and shaped the former. By imbricating them, Lacan shifted emphasis of the mirror stage from its role in the production of the imaginary to the analysis of that imaginary in the context of the adult patient's symbolic world.[104] Lacan's most famous presentation of the relationship was the *L* schema (named for its resemblance to the Greek letter lambda, λ); he introduced the schema in a seminar on Edgar Allan Poe's story "The Purloined Letter" and then chose the essay version to open his 1966 collection *Écrits* (fig. 5.1).

The schema had the imaginary built into it, and Lacan explicitly linked the two middle terms, *a* and *a′*, to "the couple involved in reciprocal imaginary objectification that I have brought out in 'The Mirror Stage.'"[105] The other vertices referred to the big Other (*A*) and subject (*Es, S*) whose relation was blocked by the axis of the imaginary. In other words, the identification of the ego (*a*) with the specular image (*a′*) got in the way of the symbolic discourse of the Other and the subject, impeding the "full speech" of the unconscious with the "empty speech" of the ego with its imaginary and alienating desires. As he wrote in the Rome Report, psychoanalysis had to reject the "idea that the subject's ego is identical to the presence that is speaking to you."[106] Only by distinguishing between the two did it become clear that it was the role of the psychoanalyst to help the patient loosen the identifications out of which the ego was constructed.

If the symbolic took pride of place in Lacan's work only from the mid-1950s onward, we can see the roots of it much earlier. As some commentators have noted, the concept was at work in Lacan's writings from the 1930s.[107] Jacques-Alain Miller, Lacan's student and the editor of his seminars, argued that in "Les complexes familiaux" the "cultural" was "an ersatz symbolic" in the sense that it was, quoting Lacan's original, "subversive to all fixity of instinct."[108] The sense of lack that drove the child to identify with its image also propelled it

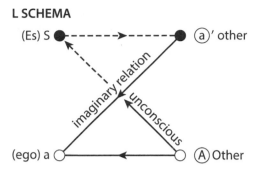

L SCHEMA

FIGURE 5.1. Lacan's *L* schema. The identification of the ego (*a*) with the mirror image (*a'*) blocked the symbolic discourse of the Other (*A*, for *Autre*) and the subject *S*. From Jacques Lacan, "Seminar on 'The Purloined Letter,'" in *Écrits: The First Complete Edition in English*, trans. Bruce Fink (New York: W. W. Norton, 2006), 6–48, on 40. Available on commons.wikimedia .org, "Schéma L." Copyright © 1996, 1970, 1971, 1999 by Éditions du Seuil. English translation copyright 2006, 2002 by W. W. Norton and Company, Inc. Used by permission of W. W. Norton and Company, Inc.

away from instinctive behaviors. In contrast to the order of animals, where instincts were fixed, the realm of the human (culture) was characterized by a variability of relationships. As Miller pointed out, this was still a long way from Lacan's later formulations. And yet, "if there is a major point in this text, and also entirely decisive after this division of the subject, it is the denunciation of the concept of instinct regarding man, the instinct as rigid, invariable, to which one opposes, precisely by cultural inquiry the most elementary, the infinite variations of human existence and its modes of organization."[109]

There is another way we can identify a protosymbolic in the earlier text, one that went back to Wallon's use of the term. As I argued, simple identification with one's reflection was insufficient for Lacan's reading of the mirror stage, just as it was for Wallon.[110] He also needed the child to overcome an earlier "animism," in which it happily accepted a double location in space. Only thus could the alienating unity of the mirror image "intrude" on the mind and be used to order the child's proprioceptive experience. As we saw, it was precisely this overcoming that Wallon had identified as the gateway of the symbolic. That is, while Lacan shifted the emergence of mirror recognition back two months in order to make his argument about anticipatory unification, he had to surreptitiously engineer an even greater shift in the emergence of the symbolic, relying on it as a condition of the possibility of the mirror stage six months before it first emerged according to Wallon.

This reading of the origins of the symbolic is confirmed by the curious re-lationship between the symbolic and the imaginary in the later work. As in the 1938 paper, the symbolic turned out to be both posterior to and a surreptitious condition of the mirror stage. Lacan did not, to my knowledge, date the mo-ment a child was swept up into the symbolic. Nevertheless, he did link it to two different developments that could themselves be dated. First, on a number of occasions, Lacan linked the symbolic to the fort-da game. This found its canonical presentation in the 1920 book *Beyond the Pleasure Principle*: Freud described his eighteen-month-old grandchild enacting the leaving and reap-pearance of the mother using a "wooden reel with a piece of string tied round it." Holding the string, the child threw "it over the edge of his curtained cot" and then retrieved it by pulling the string. The name derived from the sounds the child used: "o-o-o-o" (interpreted by Freud to mean *fort*—away) while it went away and "da" (there) when it came back.[111] By replacing the object of desire with the sign, in Lacan's telling, the child cancelled "its natural prop-erty" and thus manifested "the determination that the human animal receives from the symbolic order."[112] In the Rome Report, Lacan was even clearer: "From this articulated couple of presence and absence . . . a language's [*langue*] world of meaning is born, in which the world of things will situate itself."[113]

The second anchoring was even later, associated by Lacan with the fourth year of life: the Oedipus complex. Referring to the work of the anthropologist Claude Lévi-Strauss, Lacan related the symbolic consistently to the law of kinship that was "imperative for the group in its forms, but unconscious in its structure"—meaning that despite our pretention to freedom of choice, it de-termined our actions.[114] On this basis, Lacan referred to the "nom du père"—the name of the father—that is homophonic in French with the "no of the father." Our sense of our freedom was actually underwritten by a symbolic law that comes from the Other; this law was prototypically the incest taboo, which caused us to repress our desire for the mother. Through the Oedipus complex, the "non du père" was internalized as the superego.

And yet despite this presentation of the symbolic as subsequent to and building on the mirror stage, Lacan also implied that it must have been present from the very beginning, drawing out the implications of the 1938 paper that I laid out earlier. In his 1953–54 seminar, Lacan dedicated a number of sessions to the "topic of the imaginary." In these he described a second mirror arrange-ment, which he presented as a "substitute for the mirror stage." This apparatus centered on the inverted bouquet, first described by the physicist Henri Bouasse in his 1917 book *Optique géométrique élémentaire* (fig. 5.2).[115]

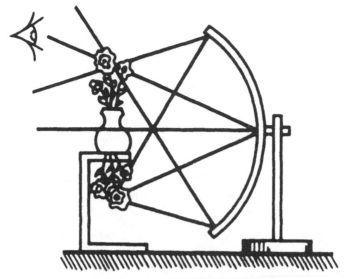

The experiment of the inverted bouquet

FIGURE 5.2. Lacan's inverted bouquet. The reflection of the flowers (in reality hanging beneath the stand) in the concave mirror lined up with the vase only when seen from a particular perspective. From Jacques Lacan, *The Seminar of Jacques Lacan*, ed. Jacques-Alain Miller, book 1, *Freud's Papers on Technique, 1953–1954*, trans. John Forrester (1975; New York: W. W. Norton, 1988), 78. Copyright © 1975 by Les Éditions du Seuil. English translation © 1988 by Cambridge University Press. Used by permission of W. W. Norton and Company, Inc.

The unity of the body in the mirror was like the vase, which appeared to contain, and thus present as coherent and unified in an ego, the multiple desires of the subject—here the bouquet, hanging beneath the stand.[116] Lacan argued that he was describing not so much a "moment of development" but its "exemplary function," in parallel with the *L* schema.[117] Yet it is clear from the diagram that the imaginary containment of the flowers, the wrapping up of the divided self by the imaginary ego, occurred only when the observer occupied a particular position. Put the eye somewhere else and the vase and the mirror image of the flowers did not line up. Lacan wrote, "In the relation of the imaginary and the real, and in the constitution of the world such as results from it, everything depends on the position of the subject. And the position of the subject—you should know, I've been repeating it for long enough—is essentially characterized by its place in the symbolic world, in other words in the world of speech."[118] That is, the symbolic provided the structure that allowed the imaginary ego to appear.[119]

Lacan was primarily interested here in the relationship between the imaginary and the symbolic in his adult patients, but it is clear that the argument held even if talking about the initial moment of mirror identification. Lacan made this connection explicitly in his tenth seminar, in 1962–63. In response to an inquiry he received about the relation between the two registers, Lacan claimed the following: "I don't believe there have ever been two phases to what I've taught, one phase that would supposedly be focused on the mirror stage and the imaginary, and then afterwards, at that moment of our history that is marked by the Rome Report, my supposedly sudden discovery of the signifier."[120] To explain the relationship, Lacan went back to the mirror encounter: The child "turns round, I noted, to the one supporting him who's there behind him." "This movement is so frequent, constant I'd say," he specified, "that each and every one of you may have some recollection of it." The child, so Lacan explained, "seems to be asking the one supporting him, and who here represents the big Other, to ratify the value of this image." That is, it was the parent who confirmed that the mirror image was actually the child: *Look it's you!* For Lacan this was an "indication concerning the inaugural nexus between this relation to the big Other and the advent of the function of the specular image, noted by $i(a)$."[121] The idea is that the child is held in a place to experience the imaginary image through the agency of a parent who is already folded into the symbolic world.

Lacan introduced these remarks with the question: "Haven't I always insisted on the movement that the infant makes?"[122] As it happens, the answer to this rhetorical question was no. At least in his published accounts of the mirror stage, he had never described the child turning back to look at its parent. If we recall, in the 1949 paper, Lacan had added reference to "some prop, human or artificial (what, in France, we call a *trotte-bébé* [a sort of a walker])."[123] A parent was thus evoked, but as a possible and not necessary part of the scene. At no point did he suggest the child sought confirmation from the adult. Seemingly recognizing this, Lacan referred in the 1962–63 seminar not to any of the canonical descriptions of the mirror stage but rather to another paper, the "Presentation on Psychical Causality," first given in 1946 and later included in *Écrits*. This paper did connect the mirror stage to the father. It was the father, Lacan said, who through the Oedipus complex made the "world of persons 'flocculate' in the subject."[124] By this, Lacan meant that the father helped precipitate a range of distinct individuals out of the desirous identification that characterized the child's relationship with its mother and sibling rivals. But as we saw in Lacan's "Complexes familiaux," he had placed the Oedipus complex as subsequent to and dependent on the mirror stage. Lacan admitted as much in the 1946 paper. In moving from the appearance of the father to the mirror stage, Lacan suggested that he was moving back to an earlier and less socially contingent development (for he was clear that the Oedipus complex was restricted to societies with the patriarchal form of the family).[125] It was not Lacan who had made the child's turn to the parent the central gesture of the symbolic. That was Wallon.

Conclusion

In his work on the mirror, Lacan came close to admitting that he did not perform the test. After all, he was a psychoanalyst, whose tool was language; in his clinical work, he simply didn't have anything to do with prelinguistic children. He even claimed that this was a moment of development into which psychoanalysis cannot penetrate.[126] As he wrote in "Some Reflections on the Ego," preverbal experience had to exist "in a form in which there was at least the possibility of its being verbalized," and thus could only be accessed through an older patient.[127] Even in the canonical 1949 paper, he emphasized the way in which he analyzed the mirror experience "based on a language technique. . . . This is why I have sought, in the present hypothesis grounded in a confluence of objective data, a *method of symbolic reduction* as my guiding grid."[128] Across his career, every account of the mirror stage was illustrated by examples of (adult) dreams of dismembered bodies.[129]

This was not simply an obstacle to be overcome. As we have seen, Lacan was most interested in the way the (adult) patient "verbalizes" past events in the *hic et nunc*. In his words, "What is at stake is not reality, but truth, because the effect of full speech is to reorder past contingencies by conferring on them the sense of necessities to come."[130] That was why Lacan criticized Freud for focusing on the historical reality of traumatic events, recovered through a process of remembering.[131] According to Lacan, it was "impossible to situate, through the anamnesis, the exact time and place at which certain intuitions, memory illusions, convictive resentments, and imaginary objectifications occurred."[132] Correlatively, the mirror stage was important for the way in which it impacted the mental life of an individual in the present, not as a moment in that individual's life history.

And yet, we should not forget that however unconcerned he was with timing more generally, Lacan did pinpoint the beginning (and end) of the mirror stage. As I have shown in this chapter, that dating was not an indifferent element in his analysis. For if the mirror stage was an "ontological operation," it was important to Lacan's argument that it was *also* a particular moment of human development. In relating the mirror stage to the "human prematurity at birth," Lacan forced himself to locate it in early childhood. We wouldn't feel the same desire for our reflections, according to Lacan's logic, if we only came to identify with them once we were fully mobile and in control of our bodies. By pushing the mirror stage back in time—further than even Wallon—Lacan opened up a crucial gap, an incompatibility between the child's lived experience and its mirror image. It was this that allowed it to be a vehicle for alienating identifications.

This difference, which made mirror recognition a form of misrecognition, was also what makes the mirror stage productive. Previously the mirror test was a tool for revealing preexisting cognitive powers. Even Gallup's emphasis on the changing nature of the mirror stimulus tracked the changing way the ape saw its reflection, not the (relatively stable) way it saw itself. For Lacan, in contrast, the mirror stage engendered a deep-seated transformation because it was always out of step with the real body. Difference was productive. In emphasizing this difference and its productivity, Lacan laid out the conceptual and practical ground on which would grow a new branch of the mirror-test tradition.

6

There Are No Mirrors
in New Guinea

EDMUND S. CARPENTER AND
THE QUESTION OF "TRIBAL MAN"

IN THE REMOTE HIGHLANDS of Papua New Guinea, a group of men and women from the Biami tribe gathered around a small hand mirror. A first look into the glass had a dramatic effect. They "cover their mouths, their eyes, hide their heads, run away, gasp, tremble, their eyes bright, hurrying back to dare a second look."[1] After this initial commotion, the Biami continued to inspect the mirror, which "paralyzed" them: they "stood transfixed, staring at their images, only their stomach muscles betraying great tension."[2]

The person confronting the Biami with mirrors and noting down their behavior was the American visual anthropologist Edmund ("Ted") Carpenter. What attracted Carpenter to the Biami was their unique relationship with reflecting surfaces. As Carpenter pointed out, buying into a decades-old trope of the "untouched-ness" of colonized societies, the Biami were "totally innocent of mirrors." The encounter with the hand mirror was thus a revelation, and Carpenter cast it in rather grandiose terms: the Biami's response showed the "tribal terror of self-awareness," as he titled his major 1975 essay.

Though trafficking in long-held stereotypes about "primitive societies," Carpenter did not seek to confirm to himself or to others the superiority of the West. Rather, in studying the response of the Biami to the mirror, he wanted to invert the hierarchies that had motivated the mirror-test tradition, and had justified colonial expansion, not least that of Australia into New Guinea. In 1973, unaware of Carpenter's research, René Zazzo had posited as a thought experiment "a man who had always been without a mirror, without cultural references to reflective surfaces."[3] He thought he could use this to separate self-recognition from consciousness: Who would think that "the ignorance of optical reflection [*reflet*] would deprive him of self-reflection [*réflexion*]?"[4] Carpenter didn't see the mirror experience as the progenitor of self-consciousness. But he did, like Jacques Lacan—despite the huge distance, both geographically and disciplinarily, between their projects—see in the desirous identification with the mirror image a terrible misrecognition, with far-reaching personal and social consequences.

Carpenter in Papua New Guinea

Carpenter arrived in Papua in the summer of 1969, after having been hired by the Australian government as a research professor at the newly founded University of Papua New Guinea in Port Moresby. He was to act as a communications consultant for the state Department of Information and Educational Services (DIES), advising on the uses of radio, film, and television "to reach not only townspeople, but those isolated in swamps & mountain valleys & outer islands."[5] Papua New Guinea had been under Australian rule since the early part of the century, remaining so until independence in 1975, and in the 1960s DIES embarked on a number of initiatives that were intended to assimilate the various Papuan populations to Australian cultural norms. It organized public showings of films, using the much advertised "cinecanoe"—a boat with a cinema on board that traveled from village to village—and involved indigenous people in radio broadcasting. Nevertheless, the Australians worried that the use of these media was counterproductive. The inhabitants of the island cities of Rabaul and Kieta, who had been in "contact with westerners since 1885," were now "marching in the streets," and a popular explanation was the recent radio broadcasts by a group of Australian administrators—Protestant sermons usually given in English.[6] Carpenter was tasked with assessing the consequences of media exposure and developing strategies for overcoming any deleterious effects.

He was the ideal man for the job. Although Carpenter often found himself at odds with the scholarly mainstream of his field, he could boast a range of impressive and relevant credentials. He was an expert on the life and art of the Canadian arctic peoples, having lived with the Aivilingmiut (a.k.a. Aivilik) for several years.[7] Up until 1967, two years before being hired by the Australians, he had served as the chair of the anthropology department at San Fernando Valley State University in Northridge, California, heading a generously funded academic program that combined experimental anthropology, filmmaking, and jazz.[8] He was also a former radio and television broadcaster and had ample experience with ethnographic filmmaking. Most importantly, perhaps, Carpenter had a powerful personal ally, having cultivated a lifelong collaboration and close friendship with the media guru Marshall McLuhan, who, thanks to the publication of his *Understanding Media* in 1964, was at the height of his fame. The two men had formed the so-called Toronto school of communication, where they cotaught a seminar on media and communication, and published their artistic journal *Explorations*, both of which were supported by a grant from the Ford Foundation.[9] In short, Carpenter was a well-known anthropologist who had both a scholarly interest in and a practical experience with media.

He also happened to be looking for a job. In 1967, in large part owing to political pressures, the funding for his program at Northridge was cancelled. Over the next two years, Carpenter lived a peripatetic life, with stints at Fordham (sharing an appointment with McLuhan) and Santa Cruz. The offer from the other side of the world was thus a lifeline, rescuing Carpenter from a precarious situation. But it was perhaps the intellectual opportunity that Carpenter found most exciting. Later he would recall that the job presented him with the unique prospect to "step in and out of 10,000 years of media history."[10]

Fall and Redemption of Literate Man

Ten thousand years was a long time, even by the standards of the Toronto school, which had fashioned an overarching narrative of human history from the advent of writing to the present. In its canonical form, presented by McLuhan in his 1962 *Gutenberg Galaxy*, the media history of mankind had begun with the introduction of phonetic literacy in ancient Greece (ca. 800 BCE).[11] The Greek alphabet introduced a new sensory regime, which ended the stage of "oral tribal culture." For McLuhan, the phonetic alphabet produced an "abstraction of meaning from sound and the translation of sound into a visual code."[12] Displacing the sphere of meaning from the "audible-tactile world into a visual world," the result was an imbalance, where the visual dominated the other senses.[13] The ramifications of this shift, McLuhan and Carpenter thought, reached the very foundations of social life:

> Beginning with the phonetic alphabet & the Greeks, there came a habit of detachment & noninvolvement, a kind of uncooperative gesture toward the universe. From this refusal to be involved in the world he lived in, literate man derived detachment and objectivity. He became alienated from his

environment, even from his body. He believed there was an elegance in detachment. He valued the isolated, delimited self, especially the mind. He became an island, complete unto himself.[14]

The advent of printing—opening the doors to the next historical epoch, the Gutenberg Galaxy—consolidated these changes, allowing the development of the modern state and religion: democracy, Protestantism, capitalism, and nationalism. In fact, McLuhan argued, Gutenberg's invention offered a much better explanation for modernity than that provided by previous historians.[15]

It is important to realize that for McLuhan, as for Carpenter, this was not an ineluctable declension narrative.[16] In his 1951 *The Mechanical Bride: Folklore of Industrial Man*, McLuhan had criticized the media of popular culture—advertising, newspapers, and comics—but he took his inspiration from Edgar Allan Poe's sailor who "saved himself by studying the action of the whirlpool and by co-operating with it." As McLuhan continued, his book "likewise makes few attempts to attack the very considerable currents and pressures set up around us today by the mechanical agencies of the press, radio, movies, and advertising." He hoped that by analyzing the effects these media could have on society, he could "get inside the collective public mind," and seek new solutions to its ills.[17]

This is why in McLuhan's later work, and in Carpenter's work at the same time, "electric" or "electronic" media was presented in more positive terms.[18] They argued that electronic media allowed a partial return to the preliterate state, mitigating the effects of the Fall. And they suggested that electronic media provided the means to preserve those societies that had managed to avoid the catastrophe of literacy. Television bore for McLuhan the greatest hopes and expectations. It offered, he thought, a unique opportunity to break with the visual regime instantiated and consolidated by literacy and print, to "retribalize" man. Television, the "most significant of the electric media" because of its pervasiveness, was "primarily responsible for ending the visual supremacy that characterized all mechanical technology."[19] Despite appearances, McLuhan thought that TV was primarily a "tactile" medium, by which he meant a medium with the "greatest interplay of all the senses."[20] It was also what he called a "cool" or "cold" medium, a medium of "low definition" that provided only little information and thus required the active participation of the viewer.[21]

Participation here was the key term, because it helped bring humanity out of the detachment and alienation that had marked the Gutenberg Galaxy. As McLuhan said, "the TV image is a mosaic mesh not only of horizontal lines but of millions of tiny dots, of which the viewer is physiologically able to pick up only fifty or sixty from which he shapes the image; thus he is constantly filling in vague and blurry images, bringing himself into in-depth involvement with the screen and acting out a constant creative dialog with the icono-scope."[22] Participation in the media foreshadowed participation in society, and thus television promised to "fulfill man's psychic and social needs at profound levels."[23] Whereas literate man was "alienated, impoverished," retribalized man "can lead a far richer and more fulfilling life."[24] An antidote to the "hot" nature of such media as writing, radio, and film, television could be redemptive.

Carpenter dealt with the salutary side of electronic media in similar ways. Whereas literacy had "ushered man into the world of divided senses," which gave primacy to vision; and thus "destroy[ing] [the] harmonic orchestration of the senses," the "electronic age" brought along its own "form of sensory pro-gramming," one that was "best known from the primitive world . . . where the senses interpenetrate & interplay, creating a sensory concert or orchestration."[25] As for McLuhan, the sensory aspects of electronic media had profound social consequences: "Where writing & print technology tore man out of the group, creating the great misery of psychic alienation, suddenly & without warning the electronic media hastened him back into the embrace of the group. Electric-ity binds the entire human community into a single tribe with much resulting erosion of individualism."[26] This was why Carpenter resisted contemporaneous efforts to limit television watching: "Television is part of the only environment today's children have ever known. To punish a child by forbidding him to watch TV is as nonsensical as threatening to deprive him of heat."[27]

McLuhan had divided film from television. The former alienated, the latter mitigated that alienation. As an anthropologist, Carpenter was more sanguine about film. Having spent several years living with the Aivilingmiut on Shugliaq (a.k.a. Southampton Island) in northern Canada, Carpenter had ample experi-ence exploring "preliterate" cultures. He had studied Aivilingmiut art and forms of living, claiming to discover there a very different regime of the senses, paired with a different idea of space and a different subjectivity and communal life. In their drawings, the Aivilingmiut employed multiple vantage points at the same time in a way that, for Carpenter, resembled the technique of the "x-ray." Similarly, when handed a photograph, the Aivilingmiut would examine it in the orientation it was given to them, conversely making fun of Carpenter's habit of turning upside-down pictures around.[28] To illustrate how their approach to the world differed from our own, Carpenter liked to use Wittgenstein's image of the rabbit-duck. According to Wittgenstein, we experienced only one at a time, either-or. But for the Aivilingmiut, this did not hold true. "They recognized rabbit-duck, not as alternatives, but as a single form," he said, "so that the rabbit was always in the duck and the duck was always in the rabbit." This principle could be found in Aivilingmiut masks as well: "Everything is there together. It's a pun in which all elements co-exist."[29]

Carpenter worried that this different sensory regime was on the brink of extinction. He reported with despair on the ways Aivilingmiut children were taught linear perspective, the "artistic counterpart of the modern notion of individualism, every element being now related to the unique point of view of the individual at a given moment."[30] Similarly, he noted how Aivilingmiut art was changed by the introduction of the souvenir trade: "We let the Eskimo know what we like, then congratulate them on their successful imitation of us."[31] Unable to prevent this corruption, it was the role of anthropology to preserve records of the endangered culture, and for this Carpenter promoted the use of film. In part his enthusiasm derived from the rapidity with which film could capture a world.[32] In a 1970 advert that Carpenter preserved in his notes, the French camera company Éclair referred to an NBC special of the same year, "Patrol into the Unknown,"[33] to "a village whose inhabitants had never seen a white man and whose style of living was unaltered since the Stone Age. Nobody knew how the villagers would react; but it was obvious that, whatever they did, they would do it only once." The process of corruption placed burdens on the camera man. In this drama, there were "no retakes. No waiting for jams or threading film."[34]

In the obituary for his PhD adviser Frank Speck, a specialist on the Eastern Woodland Native American peoples at the University of Pennsylvania, Carpenter wrote: "Culture, as he experienced it, was too rich, too full, to preserve in monographs alone. Nothing, he felt, should be lost. He used both a still camera and a movie camera."[35] In particular, Carpenter had the highest regard for an early generation of "Arctic witnesses" and their use of film, lavishing particular praise on Robert Flaherty's 1922 *Nanook of the North*, which, although staged, was "wholly accurate" in capturing the realities of Quebec Inuit life, according to Carpenter.[36] The "old-time anthropologists" did not turn to film for convenience's sake. As Carpenter noted, they couldn't "just go down to an electronic store and buy things; they had to build them. They literally built their own typewriters with phonemic scripts. They got hold of old movie cameras. They filmed. They took stills. They drew. They collected objects. They wrote poems. They did plays."[37] In his own 1959 book *Eskimo*, Carpenter followed this model, including sketches, paintings, and photographs of indigenous arctic people.[38]

Carpenter's understanding of electronic media probably informed his decision to accept the offer from the Australian government. It can be difficult to work out the development of Carpenter's views on the effects of media because of the aphoristic and composite nature of his publications and interviews (where kernels of insight were arranged freely, with little attention to chronology). But it seems that, at the beginning of his stay, his enthusiasm about the effects of media assuaged any guilt he might have felt for participating in what was clearly a colonial project. He recognized that "DIES is essentially a political instrument disguised as an educational & entertainment agency."[39] And yet, as he pointed out in a letter to McLuhan and his family, the media experiments he ran under its aegis could still be seen as an anticolonial gesture: "Of course, all this absolutely terrifies the officials here, who would prefer to have natives stay put, not migrate to town and cities, but keep their heads low in the swamp. As long as natives remained isolated and the few contacts were silly sermon efforts, change was controllable, a few literate natives gradually being absorbed into the colonial economy. But these tape recorders and transistors are no respectors of colonialism and the Australians have no clue, only hear, of what's about to happen."[40]

Again on this issue, Carpenter maintained a far more positive attitude toward film than McLuhan, seeing it as a tool by which one could record "tribal man" and promote social connection. As Carpenter pointed out in an interview in the early 1970s:

When we projected movies of their neighbors, there was pandemonium. Fear has kept villages isolated, even after tribal warfare had stopped. In one village I stood on a hilltop, looking out over a forested valley, and asked about a settlement, smoke from which rose in the distance. No one had ever been there. No one knew anyone who had. No one had seen a single member of that not-too-distant village. There were only legendary accounts. Now they were *seeing* these legendary strangers and they were absolutely wild with fascination. But no fear. In one amazing stroke, fear was replaced by familiarity.[41]

"Where the Hand of Man Has Never Set Foot":
Carpenter's Studies in Reflexivity

Carpenter's early work was thus guided by the desire to study and understand "tribal man" before it was too late.[42] In all his work in the 1950s and early 1960s, he harbored doubts that he had truly found the "primitive." He continued to believe in some sort of primordial humanity, a "preliterate" world of "tribal men" who "everywhere regarded themselves as integral parts of nature," and "belonged to a seamless web of kinship and responsibility."[43] It was not clear that the tribes he had studied fit this model. The Aivilingmiut had, after all, been in contact with literate groups for many years; for this reason, he thought that they might bear traces of a previous form of existence, but not in a pure form. In the prescript to *They Became What They Beheld*—a 1970 book of which major parts had been written before his trip[44]—Carpenter wrote: "The Tribal Man who walks through these pages is composed, like the Bride of Frankenstein, of bits & pieces from many sources. He cannot be found in any nearby jungle or tundra or city, but lives in a more remote land, in company with the savages of Rousseau & Diderot."[45] For Carpenter, "the primitive" was more an ideal type than a real existence.[46] This is why the trip to New Guinea held such promise. When Carpenter described the Biami as "Stone Age" people, he wanted to highlight their radical distance from the West.[47] The Biami promised an unprecedented form of time travel, jumping metaphorically some ten thousand years into the past, far before the supposed advent of phonetic literacy in Greece. If "tribal man" did indeed exist, surely he existed here.

Once he had passed beyond the capital of Port Moresby—which reminded him of a "southern California town with air-conditioned building, supermarkets, and a drive-in theater"[48]—and had made it to the Papuan Plateau, Carpenter felt as if he were entering a different world, where the inhabitants lived in "thousands of tiny villages, ... speaking over 700 different [and mutually unintelligible] languages."[49] The great cultural and linguistic diversity impeded interaction and exchange between the different tribes, which was an asset to Carpenter's experiments because it contributed to the isolation of these people from media—perhaps, he mused, the only such isolated societies in the world.[50]

True to his Toronto school heritage, Carpenter understood this isolation in normative terms. Carpenter's archives contain a handful of government patrol reports from the early 1960s that were presumably sent to him before or shortly after his arrival to warn him of the dangers of cannibalism.[51] The reports uniformly refer to the Papuan peoples in negative terms, calling them "intractable and truculent," "openly hostile," and "suspicious."[52] Carpenter, although wary of the cannibalism, rejected these characterizations. Instead he asserted (and romanticized) Papuan superiority in terms of social life and relation with the environment. Traditional Papuan sociality, he claimed, placed the individual within "a seamless web of kinship and responsibility" where he or she was "merged with the whole society."[53] For example, when the New Guinean put on masks, this was not self-expression. Rather it allowed individuals to "divest" themselves of individualism and assume "the collective powers of the community and environment."[54] To Carpenter, the Papuans were "a constant delight, full of love of children, humor, thrust of life." In a letter to his parents-in-law, he described them as "the most joyous people we have met so far, full of endless fun."[55] So too Carpenter valued their attitude toward their surroundings: "Dancers in floral skirts & feather headdresses put on the jungle, wrapping themselves in their environment. They became one with the plants & animals."[56]

Carpenter's decision to bring mirrors on his trip in part followed common practice. Going back to at least the early twentieth century, mirrors were a common gift in so-called first contact encounters between Europeans and Papuans; in addition to being easy to transport and cheap, they were assumed to be something new and unexpected.[57] At times this novelty seems to have been a problem. When the British patrol officer Jack Hides offered a mirror to the Huli leader Puya Indane in 1935, he "jumped back from seeing his face" before returning the object.[58] Carpenter's choice was reinforced by the conditions he thought were in place for one Papuan tribe, the Biami, for whom he deemed the novelty of mirrors even more radical. As Carpenter traveled up the stretch of the Sepik River on which the Biami lived, he noted that "neither slate nor metallic surfaces exist and, for reasons I don't understand, rivers in this area fail to provide vertical reflections, though reflections of foliage can be seen at low angles. I doubt if the Biami ever before saw themselves at all clearly."[59] That is, according to Carpenter, the Biami innocence of mirrors was just a particular instance of a general innocence of all reflective surfaces. The peculiarity of these conditions led Carpenter to focus his attention on the Biami tribe as the most extreme example of isolation. If familiarity with reflections was not a universal human experience, but one that could have, at least in one case, emerged at a particular moment, then it was possible to consider mirroring as a type of media in McLuhan's sense.

Carpenter seemed unsure about how to classify the mirror within the To-ronto school media schema.[60] As reflective naivete was the most extreme marker of the Biami's innocence, Carpenter was inclined to associate the mir-ror to the medium that marked the beginning of McLuhan's story: phonetic writing. Carpenter related every change undergone by the Biami to an ana-logue from this early stage in Western history. The initial experience of identify-ing one's mirror image with one's self, to Carpenter, was "probably always traumatic." Carpenter suggested that the mirror image was an outward projec-tion of an inner, symbolic, self, which "reveals that symbolic self OUTSIDE the physical self" and thus made it "explicit, public, vulnerable."[61] This resembled the trauma inflicted by writing: "When people first encounter writing, they seem always to suffer great psychic dislocation. With speech, they hear con-sciousness, but with writing, they see it. They suddenly experience a new way of being in relation to reality."[62] The parallels allowed Carpenter to relate the experience of the Biami to similar experiences closer to home. If the Biami spoke of the loss of the soul, "we call it loss of identity because we think in those terms. . . . But it's the same phenomenon."[63] Similarly, the Biami associa-tion between the soul and the breath or voice was for Carpenter part of a universal phenomenon, from the breath of life to Ben Jonson's "Speak, that I may see Thee."[64]

While the mirror seemed to be related to writing, the way it produced and alienated the physical from the symbolic self suggested a connection to other media.[65] In this way, Carpenter noted the same reaction when he exposed the Papuans to photography and audio recordings:

> We gave each person a Polaroid shot of himself. At first, there was no un-derstanding. The photographs were black & white, flat, static, odorless—far removed from any reality they knew. They had to be taught to "read" them. I pointed to a nose in a picture, then touched the real nose, etc. . . .
>
> Recognition gradually came into the subject's face. And fear. Suddenly he covered his mouth, ducked his head & turned his body away. After this first startled response, often repeated several times, he either stood trans-fixed, staring at his image, only his stomach muscles betraying tension, or he retreated from the group, pressing his photograph against his chest, showing it to no one, slipping away to study it in solitude.[66]

Similarly, when Carpenter used a tape recorder to record their voices and play them back to them, he found that they didn't at first "recognize their own voices & shouted back, puzzled & frightened."[67] A man in Mintima in the Chimbu

Province, whom Carpenter recorded when he was singing, at first froze, staying absolutely still until, eventually, he gave Carpenter an embarrassed look.[68]

As with the mirror, Carpenter interpreted the Biami's response—they uniformly ducked their heads and covered their mouths—as an attempt to prevent the loss of their soul: "There's a sudden loss of identity they're trying to prevent."[69] In Carpenter's interpretation of the behavior, media of identification externalized the self, rendering it vulnerable and causing the experience of "instant alienation."[70] With his characteristically dramatic style, he wrote: "In one brutal movement, these villagers had been hoisted out of a Stone Age existence and transformed from tribesmen into detached individuals, lonely, frustrated, alienated. They were no longer at home in their old environment, or, for that matter, anywhere."[71]

The effects of these "media of identification," from the mirror to the photograph, came to challenge Carpenter's optimism about film. As we saw, Carpenter had previously diverged from McLuhan in thinking that film might be redemptive. McLuhan had focused primarily on the question of participation, the way in which different forms of media required different levels of engagement from the audience. Low resolution media like television required more engagement than high-resolution media like film, and multisensory media required more synthetic labor than monosensory media like radio. This engagement then corresponded to social engagement, and thus to the question of alienation. Carpenter had tended to focus on the question of alienation directly. And from this perspective, he had hoped that film might offer a counterweight to colonialism by forging social bonds between isolated tribes, as well as preserving cultures that were threatened by their contact with Western society.

In his media experiments in Papua, however, Carpenter came to see film as yet another medium of identification and thus a cause of alienation. Experiments with film were more complicated than the others. In fact, Carpenter considered it a "minor logistic feat" to capture the effects of filming the Biami. They filmed the Biami's responses to the new media, then sent the negatives back to be developed in America, and screened these same films for the Biami again, recording their reactions using infrared light and film (fig. 6.1). But according to Carpenter, it was "worth the effort": "There was absolute silence as they watched themselves, a silence broken only by whispered identification of faces on the screen."[72]

This was also visible when the Biami began to make movies or use other media of identification themselves. After their first startled reactions, "in an astonishingly short time, these villagers, including children & even a few women, were making

FIGURE 6.1. Biami watching themselves on film. The film was sent back to the United States for development and, from there, sent to Papua Guinea for screening, which was in turn recorded by photograph and film. Harald E. L. Prins and John Bishop, "Edmund Carpenter: Explorations in Media and Anthropology," *Visual Anthropology Review* 17, no. 2 (2001–2): 110–40, on 127. Photos by Adelaide de Menil © / Rock Foundation, with kind permission.

FIGURE 6.2. Papuans using a camera, 1969. Photo by Adelaide de Menil © / Rock Foundation, with kind permission.

movies themselves, taking Polaroid shots of each other, and endlessly playing with tape recorders" (fig. 6.2)[73] Some walked around wearing their photographs on their foreheads "in front of their feather headdresses."[74]

While Carpenter declared his hope that the Biami "would present [their] own culture in a fresh way, and perhaps even use the medium itself in a new way," he came to the conclusion this wasn't the case: their work turned out to be similar in style to that of Carpenter and his crew. In a faithful application of McLuhan's principle that the "medium is the message," Carpenter concluded that these films "tell more about the medium employed than about the culture background of the author or cameraman." Denying the Biami any artistic agency of their own, Carpenter insisted that their former culture existed as "no more than residue at the bottom of the barrel."[75]

Carpenter's changing approach to film also led him to reconsider television, which for McLuhan was the ultimate "cold" medium. As Carpenter wrote, "What we did in New Guinea was, on a small scale, precisely what electronic media, especially television, have done to all of us. In my opinion, television is producing a great deal of anxiety and psychic alienation in Western society. If our experiments in New Guinea help to bring us some understanding of how this happens, we may all profit."[76] The effect was to drain the optimism McLuhan and Carpenter had invested in modern electronic media. Far from being redemptive, such media were complicit in the very process of detribalization that had corrupted the West.

Carpenter returned to the Biami village six months after his first arrival, by which time it "had changed completely. Houses had been rebuilt in a new style. The men wore European clothes, carried themselves differently, acted differently. They had left the village after our visit and, for the first time, traveled outside the world they had previously known." With the introduction of the media of self-awareness, "the cohesive village had become a collection of separate, private individuals," who "wandered 'between two worlds, one dead, the other powerless to be born.'"[77] Acutely aware of the power of media to transform tribal society, he referred to the effects of his work with combination of fascination and horror: "in one fantastic SWOOOOP, they are hoisted right out of tribal life and ready to GO, cut off from the past forever, the process of alienation and individualism, which took so long in the Western world, speeded up by electricity to the point WHERE YOU CAN ACTUALLY FILM IT."[78]

These changes led Carpenter to see himself as participating in the very project of colonial assimilation that he had once thought his work would resist. In a 1980 interview with the anthropologist Bunny McBride, he expressed regret for a recommendation he made. At the time, the Australian government conducted regular censuses of the population. Carpenter thought it could be used as a form of discipline too. Instead of physical punishment or prison for cannibalism, he had suggested creating a ritual around the census, and "make it a great show."[79]

The patrol box is brought out, with a guard on either side. The sergeant unlocks all the padlocks, opens [the box], takes out this *huge* census book, and [calls forth] the interpreter. Trembling men are standing in front of him. He gets each man's name, writes it down and *shows* it to them. And he repeats it—he shows it to them. He closes and reopens the book: There's the name; it's still there. Then he takes a Polaroid shot of each man, develops it and explains it to them: "Forehead—forehead, nose, nose—nose, that's *you*." And he staples each photo into the book with each nose. And he closes the book and he opens it, and he shows he has [each name and photo].

Then he puts the book back in the box, the guards lock all the padlocks, and he says to each man: "If you ever commit cannibalism again, we have you. We won't need to come looking for you. We have you here in the book. We have your spirit."[80]

Carpenter related the logic to the English medieval Domesday Book, which was so termed, "not because it was for taxation, but because it recorded names."[81] Here a technology of identification, both the book and the photograph, could be used as a technology of colonial control.[82]

And while Carpenter had arrived in Papua New Guinea thinking he would be able to save these so-called tribal men, he came to the conclusion that he was himself responsible for their demise. After his project was concluded, Carpenter attempted to control the damage. He never submitted his report to the Australian government. Instead, he published his results in the 1972 book *Oh, What a Blow That Phantom Gave Me!*, the title of which related the modern battle against media to Don Quixote's tilting at windmills. There, Carpenter combined aphoristic insights, anecdotes, and images in ways that the anthropologist Harald Prins and the filmmaker John Bishop have called "postmodern" *avant la lettre*.[83] Carpenter thought that the book's nonlinear form was the only way to be disruptive of colonial regimes and thus he preferred it both to the official report and to remaining silent about his experiences.[84]

For the most part, however, Carpenter blamed his media for the chaos he had introduced into the Biami's lives, and he came to question the use of media by anthropologists more broadly. As his work with the Biami made clear, knowledge of electronic media could change behavior: "Comparing footage of a subject who is unaware of a camera, then aware of it—fully aware of it as an instrument for self-viewing, self-examination—is comparing different behavior, different persons."[85] Because it was impossible to film subjects without letting them know what the media was, this change was inevitable. This compounded the destructive effect of media: "We use media to destroy cultures, but we first use media to create a false record of what we are about to destroy."[86]

When Carpenter returned from his trip, he participated in the young field of visual anthropology. He attended a three-day gathering at the Smithsonian, where he met with some of the key proponents of the emerging discipline, including Margaret Mead, Alan Lomax, Jay Ruby, and Sol Worth, a group that was responsible for the establishment of the National Human Studies Film Center. In their 1975 landmark publication *Principles of Visual Anthropology*, Carpenter contributed a piece, his "Tribal Terror of Self-Awareness."[87] But, having himself experienced the impossibility of preservation without effecting change, Carpenter soon got disillusioned and took his distance from the group and the profession at large. By withdrawing from debate, he missed the field's critical turn, which would have been more closely aligned with his perspective.[88] Instead, Carpenter grew ever more nostalgic about the lost approach of a previous generation of anthropologists. He used electronic media less and less, and although Carpenter and McLuhan remained close friends, they no longer actively collaborated. McLuhan's health deteriorated in the 1970s, and he died in 1980.

The Death of Tribal Man

While Carpenter began to worry that his anthropological tools had killed off the last existing "tribal man," the experience also led him to shift his understanding of the idea dramatically.[89] The experiments had suggested that the "detachment" and objectivity that was characteristic of postliterate societies characterized any societies that had experience of media of identification, and therefore any society that was familiar with mirrors. Carpenter thus brought back the moment of the Fall from the invention of phonetic writing to the first encounter with reflective surfaces. In so doing, however, Carpenter also expelled that Fall from history. If mirror innocence was a necessary condition for "tribal man," then "tribal man" could exist only in the peculiar environment of a small part of New Guinea, where water did not seem to reflect; it could no longer be considered a common past, shared by all peoples. After all, at what stage had the ancestors of European or indeed African or Asian peoples lived in conditions in which they were denied access to their reflections? For Carpenter, "tribal man" changed from being a historical stage in the development of humanity to a geographical anomaly.

This explains why the concept of "tribal man" mostly drops out of Carpenter's later work. In the final decades of his life, Carpenter concentrated his energies on producing an edition of the art historian Carl Schuster's work.[90] Schuster, who had passed away unexpectedly in 1969, left behind a massive quantity of unprocessed material, some "80,000 negatives; over a quarter of a million prints; 5,670 bibliographic references; files on motifs, publications, museums, geography, etc.; plus some 18,000 pages of detailed correspondence, written in at least 30 languages."[91] The collection included a range of different objects, including clothing, pottery, shamanic instruments, and artwork, from across the world and from all periods of human history. Over a period of twenty years, Carpenter transcribed and edited the material that Schuster had collected.

At first sight, Carpenter's method might seem continuous with his earlier understanding of "contemporary ancestors." In the introduction, he wrote how the volumes brought together "specimens from New Guinea and South America, medieval Germany and modern Africa," seeking to associate geographically and temporally distant societies. But as Carpenter emphasized, Schuster's "concern was not historical." Rather he sought the "underlying principles which govern their [these artifacts'] forms and meanings."[92] For instance, he drew parallels in the representations of cosmologies in objects as different as carpets from ancient Asia, a shaman's drum from the Altai people in contemporary Siberia, and the pantheon of Rome (120 CE).[93] Gone was the master narrative that allowed a scholar to locate contemporary societies on a relatively linear path from the primitive to the modern. Rather, all societies were inheritors of an earlier stage, and all societies were modern in different ways: "As we've seen, certain Paleolithic traditions survive in Papuan art. As we will see, other Paleolithic traditions survive in Eskimo art. But Papuans and Eskimos aren't Paleolithic peoples. Their arts preserve other traditions as well."[94] Carpenter came to dismiss those who adopted the attitude that had brought him to Papua New Guinea in the first place:

> In the world of electronic technology, we humbly encounter the primitive as avant-garde. Americans, Englishmen, Spaniards, Italians, Japanese flock to the Sepik, board palatial houseboats and, drink in hand solemnly view savages on the hoof. This search for the primitive is surely one of the most remarkable features of our age. It's as if we feared we had carried too far our experiment in rationalism, but wouldn't admit it & so we called forth other cultures in exotic & disguised forms to administer all those experiences suppressed among us. But those we have summoned are generally ill-suited by tradition & temperament to play the role of alter ego for us. So we recast them accordingly, costuming them in the missing parts of our psyches & expecting them to satisfy our secret needs.[95]

The search for "tribal man" was a foolish endeavor, which distorted our understandings of other cultures.

The change in Carpenter's thinking coincided with larger developments in anthropology at the time. The 1960s were marked by a movement known as "urgent anthropology," which, in response to the expanding reach of Western culture, sought to preserve "cultural isolates," often with the use of still or motion pictures, before it was too late.[96] And yet, by the 1970s this view had come under fire for both political and epistemological reasons. In the wake of widespread decolonization, anthropologists began to think more critically about how their work had contributed to the colonial project, a self-reflection that seemed only more necessary after the revelation that anthropologists had been involved in government-run and government-funded programs as part of the Cold War; these included Project Camelot, the US government's counterinsurgency research program in South America.[97] So too, as part of a larger response in the social sciences against forms of positivism (the *Positivismusstreit*—quarrel over positivism—in which the Frankfurt school social theorist Theodor Adorno played a major role), anthropologists began to question the scholar's ability to describe society in disinterested terms. This skepticism over positivism increasingly informed the debates over preservation, for it undercut the idea that anthropologists could grasp the "primitive" without changing it: the very instrument of recording was found to have corrupting effects. From the early 1970s onward, anthropologists dispensed with the idea of the camera as an objective recording device and advocated for a greater sophistication in the use of visual media, which involved training in the field of visual communication.[98]

Conclusion

Carpenter's test represents the paradoxical pinnacle of the mirror-test tradition. On the one hand, it is the fulfillment of its earlier hopes. As we saw, the mirror test developed in response to the problem that we were unable to examine human prehistory directly. That is why people like Tiedemann turned to the baby diary; the development of the individual in the first years of life acted as a proxy for a social transformation that had been lost to time. For a short moment, Carpenter thought that he was able to observe this social transformation directly. According to the spatialized logic of time that so often undergirds anthropological practice, Carpenter had discovered what he took to be the beginnings of humanity, alive and well in the highlands of Papua New Guinea, and available for study.

On the other hand, the ambiguity of the mirror test led Carpenter to think that the self recognized in the mirror was a fallen one. What had once been taken as a sign of human superiority had become a symbol of corruption. In this he came close to the claims of the French anthropologist Claude Lévi-Strauss, most famously in his "Writing Lesson" from 1955. There, Lévi-Strauss told of his encounter with the Nambikwara of Brazil, whose leader adopted the practice of writing after seeing Lévi-Strauss scribble in his notebook.[99] The leader used his newly acquired skill to consolidate his social position, which led Lévi-Strauss to conclude that the effects of writing were less of an intellectual than of a social nature. Writing was not the sign of an advancing civilization, but the first sign of corruption in a pristine society.[100]

When he was writing, Lévi-Strauss could still embrace the notion that non-Western peoples were our "contemporary ancestors." In fact, he was one of the major proponents of "urgent anthropology." In a 1966 programmatic article, he laid out his position:

> Native cultures are disintegrating faster than radioactive bodies; and the Moon, Mars, and Venus will still be at the same distance from the Earth when that mirror which other civilizations still hold up to us will have so receded from our eyes that, however costly and elaborate the instruments at our disposal, we may never again be able to recognize and study this image of ourselves.[101]

As Lévi-Strauss's plea makes clear, however, such preservation was necessary in order to preserve a privileged site for the investigation of the Western self, using the Other as a mirror to reflect on ourselves.

Lévi-Strauss's argument was prominently challenged by the philosopher Jacques Derrida in a seminar he taught in 1966, and later published in *Of Grammatology*, where he offered a close reading of "A Writing Lesson."[102] Derrida commented on Lévi-Strauss's "Rousseauist" reading of the encounter, which posited the corruption of a "crystalline" society through the encounter of writing.[103] He agreed with Lévi-Strauss that writing could establish power differences, but he suggested that we needed a broader view of it, one that went beyond Lévi-Strauss's Western ethnocentric account. If we were no longer guided by what Westerners considered writing, Derrida argued, but rather by a generalized version—what Derrida called "arche-writing"—it would become clear that the Nambikwara were not bereft of writing, and thus social hierarchies, before Lévi-Strauss's arrival.[104] Lévi-Strauss inadvertently pointed to the existence of this broader form of writing when he described the way the Nambikwara added "dots and zigzags on their calabashes."[105] It followed then that we could not see the Nambikwara as innocent, waiting for a singular moment of corruption. Nostalgic accounts of an innocent and preliterate society were ethnocentric myths.

In Carpenter's engagement with the mirror in Papua New Guinea, we see a similar development. Carpenter too expressed anxiety that the so-called primitive societies he studied might not be truly pure, and his investigations with the mirror led him to the conclusion that there was no age before media, no prelapsarian moment. Mirrors were everywhere, and so human history could not be told as a story of a fall from "tribal man" and a future redemption from the thralls of writing, but rather as the multiple and contradictory effects of many different forms of media. But this merely sharpens the irony of Carpenter's trip to Papua New Guinea: he came to this conclusion—the realization that "tribal man," understood as humanity before mirrors, was a myth—only when he found a living specimen.

7

Diseases of the Body Image
and the Ambiguous Mirror

A YOUNG WOMAN, wearing only her underwear, is standing in front of a body-sized mirror. She looks at the image, scanning it, examining body part by body part, and describing what she sees. A therapist is sitting nearby, commenting on each step of the description:

> THERAPIST: I'd like to ask you to describe yourself from head to toe. I encourage you to use as neutral statements as possible. Every time when you make a negative statement, I will interrupt you by coughing slightly so that you can correct your statement. Please describe your head first. What do you see when you look into the mirror?
>
> MS. S.: So, first I notice my awfully large nose.
>
> THERAPIST: Hmhhmmm! That is a very negative description. Try again.
>
> MS. S.: Yes . . . well . . . I see a straight nose that is not snub-nosed as I would like it to be.
>
> THERAPIST: Hmmhmmmhmm!!
>
> MS. S.: This is very difficult. Let's see. I have an oval face with a straight, even Roman nose . . . that is very slim and tapers at the end, and the cartilage protrudes a little at the top.
>
> THERAPIST: Yes, very good. Please continue . . . [1]

In this example, we are witnessing the struggle of an anorexic patient, assisted by her therapist, with her mirror image. The mirror plays a curiously ambivalent role in the process. It is the cause for the patient's frustrations: it puts in plain view the unloved parts of her body—her "awfully large nose." But it also holds out the promise for therapy, guided by the idea that once the patient manages to accept the mirror image in a particular way—as an external

corrective to her inner experience of her body—she is on the path to improvement. The therapist is present to encourage the patient to describe her reflection in the mirror in neutral terms, that is, as *it really is*.

Here, the mirror serves a different purpose than it did before. In earlier
chapters, we saw that it primarily had a diagnostic capacity, the ability to detect
a particular mental development. Now it has a therapeutic purpose. The difference, however, isn't quite as great as it first appears. As we saw, especially
with Lacan and Carpenter, several scientists noted the transformative role of
the mirror encounter. The mirror experience was involved in the emergence
of self-awareness. Of course for them, the egoism produced by the mirror was
something that needed to be countered, its deleterious effects mitigated. But
in foregrounding the productive power of the mirror, and exploiting the space
between the mirror image as a vehicle for misrecognition and as an objective
reflection of the body, a range of researchers also came to see it as a means for
positive change, a tool for reshaping feelings and perceptions that were otherwise beyond the reach of medicine.

The emergence of the mirror as a tool to treat disorders such as anorexia
was the result of a long development in the understanding of the disease.
The dominant approach, developed by the émigré psychiatrist Hilde Bruch
(1904–84), in the early postwar period, presents anorexia as a disturbance of
the body image, a distortion of how we feel and live in our bodies. This insight
was originally couched in the language of psychoanalysis. Consequently,
Bruch placed her therapeutic hopes on a working through of familial relations,
most importantly the relationship of her patients with their mothers. But as
Bruch's theories came under attack by feminist thinkers in the 1970s and 1980s,
combined with the general decline in fortunes of psychoanalysis, physicians and
psychologists set their sights on remolding the body image directly. The mirror
turned into a powerful tool for grappling with this otherwise elusive beast.

Hilde Bruch and the Body Image

Anorexia, or anorexia nervosa, produces startling changes to the physical
body. Young healthy people, most often women, transform from rosy vitality
to emaciated shadows of their former selves, the famous "skeleton only clad in
skin."[2] The physicality of the disease is clearest when it leads to death, such as
in the case of the American pop singer Karen Carpenter. Before she died in
1983 of heart failure associated with low serum potassium caused by starvation,
critics had commented on her increasingly emaciated appearance.[3] Afterward,

the dangers of anorexia were propelled into public consciousness.[4] The shocking physicality of the disease was accentuated by another, seemingly contradictory, aspect: the radical *denial* of bodily needs, an asceticism suggesting victory of the mind over the flesh. It is perhaps this strange relationship that has made people repeatedly refer to the disease as "bizarre," or as an "enigma."

In the 1970s and 1980s, historians tended to play up this tension between physical and immaterial factors. In her classic history, Joan Jacobs Brumberg placed anorexia nervosa in a far longer tradition of the "refusal of food," from the fasting of female saints in the Middle Ages to the so-called fasting girls of the Victorian era and the present—a "transition from sainthood to patienthood."[5] Across these different contexts, Brumberg noted a persistent clash between male rationality and feminine challenges to it, leading to a phenomenon that strained against the somatic limits imposed by male scientists.

The most important modern theorist of the disease, Hilde Bruch, took this tension as her starting point. As she pointed out in her book *The Golden Cage. The Enigma of Anorexia Nervosa* (1978), despite the "weakness associated with such a severe weight loss, they [anorexics] will drive themselves to unbelievable feats to demonstrate that they live by the ideal of 'mind over body.'"[6] Anorexia was a "puzzling disease, full of contradictions and paradoxes."[7]

To account for these contradictions and paradoxes, Bruch sought the mechanism for anorexia's symptomology in the "body image." The advantage of such a concept was that it provided her with a quasi-somatic explanation—the body image was assumed to have a neurological, that is, physical, foundation—that also paid attention to psychology; the body image structured our *experience* of our bodies. By foregrounding the body image (at least in her earlier work),[8] Bruch set the frame for almost all research into anorexia thereafter; in the most recent version of the *DSM*, the condition is still defined in these terms.[9]

The body-image concept has its origins in the somatic tradition leading back to British neurologists Henry Head and Gordon Holmes. In 1911 they described it as a physical structure, providing a postural model of the body. It served as a kind of reference point or schema, itself "remain[ing] outside central consciousness."[10] Against it all changes of body movement and positions were measured, with the body image constantly changing as well: "Anything which participates in the conscious movement of our bodies is added to the model of ourselves and becomes part of these schemata."[11]

This idea was picked up and expanded by the Austrian American psychiatrist and psychoanalyst Paul Schilder. Schilder defined the body image as "the picture of our own body which we form in our mind, that is to say the way in which the body appears to ourselves."[12] Schilder brought the body image into conversation with a tradition that has appeared before in our story: the associationism of the Germanophone neuropsychiatric tradition.[13] The body image provided the intuition that our body was unified, and "although it [this unity] has come through the senses, it is not a mere perception."[14] Rather, drawing on Head, Schilder argued that it was a schema, born of nervous associations, for interpreting incoming sensory impressions. For example, when receiving a tactile sensation and at the same time seeing one's left hand being touched, one learned to associate the sensation with that location. When the body schema was functioning correctly, the same sensation would be on future occasions experienced as sensation on the hand. The body schema was thus created from past sensory impressions that helped organize and constitute new incoming ones. As Schilder wrote, the impressions "form organized models of ourselves, which can be called schemata; these schemata change the impressions that come from sensibility such that the final sensation of the position or the place enters consciousness already in relation to past impressions."[15] It also took into account motility, in the broadest sense (Schilder included personality and emotions), as well as social and libidinal aspects, a broad scope that Bruch termed "holistic."

Bruch considered the "body image" to be a notoriously difficult concept, an assessment that is shared by the history of writings on body image in anorexia, and perhaps about the body image more broadly.[16] She spent a lifetime grappling with it. As a glance through her notes and correspondence from the length of her career reveals, she was preoccupied and never fully satisfied with the state of knowledge of the concept and its clinical implications. In late 1965, she co-organized the two-day "Symposium on Research in Body Image" in Houston.[17] She invited researchers writing on related topics and conditions—such as psychoanalytic concepts of body perception, the relationship between object perception and body perception, and conceptual and methodological problems in body-size judgment research—and took detailed notes.[18]

Even with the concept's difficulty, Bruch placed the body image at the heart of her anorexia research. "The first symptom [of anorexia]," she argued, "is a disturbance in body image of delusional proportions."[19] Because of this disturbance, patients perceived themselves to be bigger than they really were. They denied the wasted state of their bodies and showed no concern about their emaciation.[20] In addition, and this is where she went beyond Schilder, Bruch stressed the importance of enteroceptive stimuli, such as hunger: "I came to the conclusion that the correct or incorrect interpretation of enteroceptive stimuli and the sense of control over and ownership of the body needed to be included in the concept of body awareness or body identity."[21]

Psychoanalysis and Mothers

Bruch framed her understanding of body-image distortion through an appeal to psychoanalysis.[22] Sharing the fate of many Jewish physicians in 1920 and 1930s Germany, Bruch fled the growing anti-Semitism at the hospital in Leipzig where she worked and emigrated to the United States (via London) in 1934.[23] At the time, it was relatively easy for a young physician from Germany to have her license accepted, and she trained in psychiatry, first in New York at the Presbyterian Hospital, then in Baltimore, where she worked with psychiatric leaders of the day Adolf Meyer and Leo Kanner. Baltimore proved particularly influential for her, as there she got in touch with the Washington-Baltimore psychoanalysis group and met the neo-Freudian Frieda Fromm-Reichmann. Fromm-Reichmann, an émigré like Bruch, practiced at Chestnut Lodge in Rockville, Maryland.[24] She invited Bruch to participate in her weekly seminars in psychoanalysis and encouraged her to train as a psychoanalyst under her guidance.[25]

As much merit as Bruch found in the psychoanalytic theory of trauma, which emphasized the importance of past experience for mental life, she also found the field intellectually stifling. She worried that "independent questions, particularly about the validity of underlying theoretical assumptions, were not only not welcomed but were immediately branded as indicating something unfavorable about the individual who asked them."[26] For the conceptualization of eating disorders, Bruch departed from the psychoanalytic orthodoxy. Most importantly, she deemphasized the sexual element. Pace Freud, anorexia did not consist in a repudiation of sexuality, manifested in the obstruction of sexual maturation through excessive dieting. Nor was the oral element of overbearing importance. Bruch also deemphasized the role of the unconscious. What Bruch did pick up from psychoanalysis, however, was the role of family dynamics in etiology; it guided her understanding of the disease as a developmental problem.[27]

As we saw, Bruch gave enteroceptive stimuli a prominent place in her account of the body image. She considered the regulation of these stimuli to be "essential for normal development."[28] For example, if a child was crying because it felt tired, and was then put to bed, the child would come to recognize the close and beneficial connection between tiredness and sleep and thus begin to develop healthy life practices. This constituted a well-balanced situation, an "appropriate" environmental response.[29] If, on the other hand, the "regular sequence of felt need, signal, appropriate response, and felt satisfaction" was not put in place, enteroceptive stimuli would not become associated with beneficial responses, causing "disordered bodily functions and body concept," and thus a loss of autonomy.[30] Most importantly for Bruch's purposes, a similar mechanism was at work with hunger. If a mother ignored the child's desires, and fed it on a schedule and in quantities based on her own judgment, the child could be led to "feel that he does not own his body, and that he is not in control of its functions. There is an over-all lack of awareness of living one's own life, a conviction as to the ineffectiveness of all efforts and strivings."[31]

Given the importance of parents in providing the appropriate response to the child's needs, Bruch identified the main illness-causing factor in anorexia as an overbearing and dominating mother.[32] Mothers should be paying sufficient heed to "child-initiated clues," such as recognizing when the child was hungry and then feeding it; but instead, "good things were bestowed without being specifically geared to the child's own needs or desires."[33] Mother always knew best, and the process of feeding was, if anything, about herself. In many cases, the child's eating behavior would be read as a commentary on the relationship with the mother, where "noneating may be equated with criticism of the mother, and eating as expressing happiness and love."[34]

The lack of autonomy of the anorexic child showed itself outside the family as well. Patients tended to have only one friend at any given time, and with each, they would "develop different interests and a different personality. They conceive of themselves as blanks who just go along with what the friend enjoys and wants to do. The idea that they have their own individuality to contribute to a friendship never occurs to them."[35] One girl noted that she felt "full of my mother—I feel she is in me—even if she isn't there." In this last example, the mother assumed a status somewhere between food and a competing body image, a foreign structure that gave the girl shape.[36]

At no point in her writings did Bruch spell out for us how precisely the mother's "inappropriate" responses would distort the body image in ways that led the patient to misjudge his or her size. Rather Bruch focused on the regulation and interpretation of enteroceptive stimuli, the "*disturbance in the accuracy of the perception or cognitive interpretation* of stimuli arising in the body, with failure to recognize signs of nutritional need as the most pronounced deficiency."[37] According to Bruch, we could take anorexics at their word when they claimed not to be hungry, for they were unable to "recognize when they are hungry or satiated, nor do they differentiate need for food from other un-comfortable sensations and feelings." They had not learned to rely on their own bodily perceptions to decide if they were required food; rather, they "need signals coming from the outside to know when to eat and how much; their own inner awareness has not been programmed correctly."[38] The observation of a pronounced overactivity and strenuous exercising in anorexic patients was of a piece with the lack of hunger awareness; they suffered from a similar lack of awareness of their fatigue.[39]

A second aspect of the disease drew on the developmental psychologist Jean Piaget. The childhood environment of anorexic patients was not conducive to the maturation of their cognitive faculties. Bruch pointed out that, despite being generally excellent students, anorexics "continue to function with the moral convictions and style of thinking of early childhood"; this was, per Piaget, "the phase of preconceptual or concrete operations."[40] As a result of this developmental delay, they tended to "live by the rules" and were "driven to be good," without developing their own sense of autonomy.[41]

The disease seemed to offer a way out. According to Bruch, "by controlling their eating, some [anorexics] feel for the first time that there is a core to their personality and that they are in touch with their feelings."[42] Starving was a form of exaggerated action that served to assert the patient's autonomy (whereas binging was considered a shameful loss of control). Patients further realized that the refusal to eat had enormously disruptive power within family life and thus that the disease provided an "accumulation of power." This power gave them "another kind of 'weight,' the right to be recognized as an individual."[43]

Bruch's account of anorexia informed her prescriptions for treatment. She was clear that a transformation of the body image was crucial for therapy: "Without a corrective change in their body concept, improvement [in weight gain] is apt to be only temporary."[44] But, as in the initial distortion, she did not elaborate the mechanism by which the body image could be reshaped in order to give a more accurate account of the patient's actual size. Because it was the social relationships around the child that had caused the disturbance of the body image, this was where Bruch focused her attention. In fact, in Bruch's discussions of treatment, the body image itself tended to drop out of the picture. She continued to emphasize the etiological importance of early development, and it was here that her therapeutic focus remained.

First, in helping an anorexic patient, the analyst should not make the same mistake as the mother, treating the patient in an overbearing way that would not allow her to develop her own feelings. As Bruch pointed out, it was important to show the patients that "there are feelings and impulses that originate in themselves and that they can learn to recognize them."[45] For many patients, therapy "was the first consistent experience that someone *listened* to what *they had to say* and did not tell them how to feel."[46] In practice, this constituted a break with traditional psychoanalytic therapy. As Bruch pointed out, if an analyst actively sought the latent meaning behind the patient's explicit words and communicated this with the patient, they would reinforce "a basic defect in their personality structure, namely the inability to know what they themselves felt, since it has always been mother who 'knew' how they felt."[47]

For a similar reason, Bruch rejected an approach from the opposite end of the therapeutic spectrum: behavior modification. Behavior modification was based on the idea that "food refusal is a learned response that needs to be changed,"[48] which could be achieved through a system of reward and punishment. Weight gain would be "positively reinforced" whereas weight loss was discouraged by forms of forced feeding. As Bruch pointed out, this method could be enormously damaging because, by taking away control, it "increases the inner turmoil and sense of helplessness in these youngsters."[49] It did not do anything to fix what was really broken, that is, to improve the relationship between self and world by putting the patient into a healthily interactive relationship with his or her social surroundings.[50]

Instead treatment should focus on the underlying social causes of the disease. First, the weight of the anorexic person had to be restored to a healthy level—when the body was starving it produced an "almost toxic state" causing disorganized thinking and sensory perception.[51] After this, Bruch followed with a program that she called "family disengagement." The "close bond" between patient and family, especially mothers, that "interferes with [patients'] developing a sense of a separate identity" needed to be severed,[52] and a more normal relationship established.

Patient Edith was a case in point.[53] When Edith developed anorexia, her mother—who had behaved toward her daughter in a "clingy" way, not giving her the independence needed in her development—started to become depressed herself. Treatment helped the mother recognize and address the marital problems that had caused her depression, and act less possessively toward her daughter. Edith at first was resistant to these changes. But her own treatment with Bruch helped her come to terms with the "feeling that she had deserted her family, that she had been like a parent in her home, that without her presence the conflicts would come into the open and the family would collapse."[54]

Feminism: From Mother to Society

In her disease explanations, Bruch acknowledged the importance of social norms in shaping the views and behaviors of her patients and their family members. The tendency of parents and especially mothers to be overresponsive to their children's needs, for example, was a result of "young parents' fanatic belief in the theories of modern child care according to which it is most important for an infant that his needs and demands are satisfied immediately and at all times."[55] As she pointed out in a 1952 talk, "Must Mothers Be Martyrs?," when a mother respected "her own justified needs," and thus avoided "feeling like a martyr" in her total commitment to her child, it allowed her child to gain a greater sense of autonomy and was thus to the child's as much as to the mother's benefit.[56] Bruch did not, however, engage in a more specific critique of the social norms that shaped these theories, focusing her attention on the individual family situation.[57]

As for anorexia, Bruch was very clear that the disease mostly affected girls and young women (anorexia in the male was "exceedingly rare").[58] So too, she argued that the growing prevalence of anorexia in the postwar period was caused by "the enormous emphasis that Fashion places on slimness," promoted in magazines, movies, and especially, television.[59] The slimness ideal was mediated by family members, most often the mother: "A mother or older sister may communicate through her behavior or admonitions the urgency to stay slim. It is not uncommon that there is an older overweight sister or cousin in the family, and the younger child observes how much pain is provoked by being fat."[60] But it is telling that most of this discussion took place in the preface to her best-known book, *The Golden Cage*, published toward the end of her career, and in part a response to the critics to whom I will now turn.[61] There was little integration with the core of her theory.[62] In fact, Bruch's primary etiology for the disease was gender neutral (with respect to the child if not the parent). It is interesting to note that, when referring to an "infant," Bruch usually used the pronoun *he*.[63]

In the 1970s and 1980s, feminist writers such as Kim Chernin, Susie Orbach, and Mara Selvini Palazzoli exploited this gap in Bruch's account, adding to her theory a set of observations about oppressive gender ideology. Such ideology affected both parties in the mother-daughter relationship.[64] Chernin, for instance, argued that the daughter was "torn between her loyalty to her mother and her response to that new woman," the new ideal of the autonomous female that "surpasse[d] the mother."[65] A daughter felt guilty about "depleting" the mother, the provider of food, so that the symbolic meaning of eating was to hurt the mother, explaining the shame felt about eating. Starving, on the other hand, became an atonement for "the crime of matricide."[66] This explanation relied on cultural tropes about the meaning of motherhood, which for Chernin was "perpetuated by the feminine mystique," invoking Betty Friedan's famous account of the emptiness of the lives of affluent American women.[67] As a consequence, this explanation, while focusing on the mother-child dynamic, allowed Chernin to account for gender difference: for boys, the same culture of matricide took away his guilt; boys were expected to thrive on the depletion of their mothers.[68]

Similarly, for Susie Orbach the anorexic daughter hurt the mother "in the most powerful ways she knows": by rejecting her food "while simultaneously carrying out what she imagines to be her mother's wish, which is for her to disappear"; the wish stemmed from patriarchal society, where it was clear that her mother really wanted her to be a boy.[69] In line with Chernin, Orbach emphasized that self-starvation (like overeating, the main focus of her book) was a rational choice when the girl was confronted with a social dilemma, one of the "adaptations to a female role which has quite limited parameters. Both syndromes express the tension about acceptance and rejection of the constraints of femininity."[70] As suggested by her major book's title, *Fat Is a Feminist Issue*, overeating (like self-starvation) was an "attempt to break free of society's sex stereotypes."[71]

Later thinkers such as Susan Bordo and Naomi Wolf built on this work (they shared the emphasis on depathologization), but they sought to recast the child's response to these social norms. Enraged and vitalized by the Clarence Thomas hearings, where Supreme Court justice Thomas was confirmed despite credible accusations of sexual harassment, they emphasized the agency and power of women over their more passive roles in what some called the "victim feminism" of the second wave.[72] In her classic work of what came to be known as "third-wave" feminism, *The Beauty Myth* (1991), Wolf pointed out that eating disorders did not start as a disease but began "as sane and mentally healthy responses to an insane social reality"—in line with the depathologization of the second wavers.[73] Nevertheless, she disliked the way second-wave feminism focused on the dynamics of the inner family circle (the mother-daughter relationship); rather, she emphasized the public aspects of the problem: "Women do not eat or starve only in a succession of private relationships, but within a public social order that has a material vested interest in their troubles with eating."[74] Wolf described what she called the "One Stone Solution":[75] "By simply dropping the official weight one stone below most women's natural level, and redefining a woman's womanly shape as by definition 'too fat,' a wave of self-hatred swept over First World women, a reactionary psychology was perfected, and a major industry was born. It suavely countered the historical groundswell of female professional success with a mass conviction of female failure, a failure defined as implicit in womanhood itself."[76]

For all their engagement with Bruch, these feminist writers avoided the concept of the body image. In part this was because a new set of studies had suggested that the overestimation of body size through the body-image mechanism was not restricted to anorexics. Researchers found the same phenomenon in bulimia.[77] And then it was found that "overperception of body width is not restricted to populations with eating disorders" more broadly. It was found in *all* women.[78] If the pathology was so widespread, it was unclear how it could explain the specifics of anorexia.

Feminist critics rejected the body image as a path to understanding anorexia for another reason. They considered body image, or rather BID (the acronym for "body-image disturbances"), to stand for the medical approach more broadly, and for all that was wrong about current approaches to eating disorders. For them, doctors focused their therapeutic powers on the individual patient, thus occluding the impact of social factors. For example, Susan Bordo, in her influential *Unbearable Weight: Feminism, Western Culture, and the Body* (1993), urged her readers to see eating disorders "as a social formation rather than personal pathology."[79] Bordo even singled out Bruch as a "dramatic example" of the medical paradigm that, by seeing anorexia as a "visuo-spatial problem, a perceptual defect, firmly placed anorexia within a medical, mechanistic model of illness."[80] The "anorectic does not 'misperceive' her body; rather, she has learned all too well the dominant cultural standards of *how* to perceive."[81]

Cognitive Behavioral Therapy (CBT)

While Bruch's psychoanalytically informed family drama was challenged by feminists outside of medical circles, it was also being undermined within them. The story of the decline of psychoanalysis in American medicine during the second half of the twentieth century is well known.[82] One of the most potent challenges came from behavior therapists, who followed Karl Popper in considering psychoanalysis to be a prime example of unfalsifiability.[83] The British behavior therapist Hans Eysenck did much to disparage psychoanalysis and present his work as a viable alternative. Comparing the improvement of psychiatric patients who received psychoanalytically informed therapy with those who received no treatment at all, he concluded that psychotherapy was, for all intents and purposes, useless.[84]

Behavior therapy drew on the principles of behaviorism, such as the work by Ivan Pavlov, Clark Hull, and B. F. Skinner that I discussed in the first part of this book. Rejecting all appeals to introspection as excessively subjective, these therapists focused on examining and controlling external, that is, intersubjectively observable, behavior.[85] The South African psychiatrist Joseph Wolpe, for instance, produced neurotic behavior in laboratory animals experimentally, which formed the basis for his fear-reduction techniques. In his work on "reciprocal inhibition," Wolpe found that the feeding response in animals inhibited fear; when he repeated the experiment many times, he was able to condition a permanent inhibition of fear. He then extended this technique to humans, using relaxation methods instead of feeding. Through this process of conditioning, a "systematic desensitization" could be achieved.[86] At the Institute of Psychiatry at the University of London, Eysenck formed another center of behavioral treatment. He and his colleagues argued that the disturbed behavior *itself* was the problem and had to be addressed in treatment; it had to be unlearned using the methods of behaviorism such as desensitization.

Behavior therapy began to run out of steam in the 1970s. Though it worked well for certain diseases (e.g., overcoming avoidance behavior, for instance, in anxiety and phobias), it had significantly less success for others, such as depression. It was also out of line with the new emphasis on the cognitive in psychology and related fields, which I discussed in chapter 4. Behavior therapy regained momentum in the 1980s, however, when people found ways to integrate it with a cognitive approach. The result—cognitive behavioral therapy (CBT)—offered a welcome solution to some of the limitations of behavior therapy, by broadening the spectrum of diseases it could treat, and by moving beyond the dogmatic denial of the patient's inner life.

Cognitive behavioral therapy worked by building on what behavioral therapy did best. Fears and phobias could be treated by desensitization. If a patient with agoraphobia walked across open spaces often enough, he or she would build up a repository of experiences that slowly lowered the acuteness of the emotion.[87] This emotional numbing would then be stabilized through cognitive insight: open spaces were not objectively dangerous, the likelihood of bad things happening to the person crossing them was indeed very small. Since the 1990s, a new wave of CBT has emphasized the importance of acceptance and mindfulness.[88] Inspired by an eclectic set of approaches including Eastern philosophy, Western stoicism, and twentieth-century medical developments such as the humanistic psychology of Carl Rogers and others, practitioners have encouraged the nonjudgmental acceptance of thoughts and emotions in their patients, rather than attempting to change them. Whatever the status of the cognitive in its approach, it is clear that CBT is, as one practitioner has put it, about the "here and now."[89] It sidelined the close examination of the patient's personal history that had been so central to Bruch.[90] In doing so, it opened up new avenues for treatment.

Body Image to Mirror Image

As we saw, the body image was central to Bruch's work, and she discussed the distortions by which patients would experience themselves as larger than they actually were. As concerns the etiology and treatment of the disease, she focused primarily on the regulation of enteroceptive stimuli, mostly hunger. The turn to CBT, however, encouraged anorexia researchers to return to the body image and the way it structured the experience of our bodies. Perhaps CBT might be able to challenge these misperceptions directly.

Hopes were raised by the curious relationship of the body image to the mirror. The body image shaped the way in which we experience our body, leading us to misjudge our shape and size. For some of Bruch's patients, the mirror image suffered a similar distortion, reproducing the pathology and trauma of their disease. One noted: "I really cannot see how thin I am. I look into the mirror and still cannot see it; I know I am thin because when I feel myself I notice that there is nothing but bones."[91]

Yet in projecting an image of our bodies into exteroceptive space, the mirror also seemed to be able to bypass the distortions of the body image, at least temporarily. Bruch noticed that "a patient will admit to having been shocked by her cadaver-like appearance on catching an unexpected glance of herself in the mirror, failing at first to recognize herself." She clarified that "usually, within a short time, this self-critical reaction is overruled by the predominant inner picture."[92] But there was a chance that the effect could be more enduring: "After repeated self-confrontation," which Bruch specifies no further, "it became more difficult for her to maintain the denial, and a change in her body image occurred so that thinness became ugly rather than comforting to her."[93]

What was valuable about the mirror image, then, was that it occupied a liminal position with respect to the body image, both as its projection into

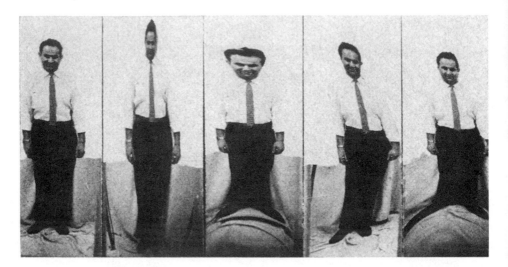

FIGURE 7.1. Four mirror distortions of the subject's body. The image on the left was undistorted. Arthur C. Traub and J. Orbach, "Psychophysical Studies of Body-Image: I. The Adjustable Body-Distorting Mirror," *Archives of General Psychiatry* 11 (1964): 53–66, on 61. Reproduced with permission from journal. Copyright © 1964 American Medical Association. All rights reserved.

space, and as an alternative to it. Later research focused on and teased out more explicitly this strange ambiguous property. A 1964 study by Arthur Traub and J. Orbach (no relation to Susie) claimed to be the first to have determined experimentally "the picture that the person has of the physical appearance of his body."[94] To do so, the researchers used an adjustable body-distorting mirror, "designed to explore the visual perception of the physical appearance of the body."[95] In the setup, the subject stood upright in front of a mirror that gave a "grossly distorted reflection of himself" (see fig. 7.1).[96] The subject was then given the possibility of adjusting the mirror image using a set of switches, until the reflection appeared correct to him, in terms of its size and shape.

The experiment reproduced the duality of the mirror image that Bruch had noted. First, it provided a means to externalize the body image, recognizing how the body "felt" to the subject. But second, it also provided, in the undistorted form, access to the body absent these changes. The diagnosis of pathology or normality of someone's body image depended on a difference between the two.[97]

The comparative element of this approach came out more clearly in another set of studies that used line drawings of the subject's own body. Consider the

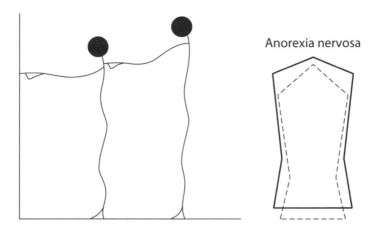

Anorexia nervosa

FIGURE 7.2. The solid line is the "subjective" drawing by the anorexic patient, the dotted line is the "objective" drawing by the experimenter standing behind the patient while both are looking into the mirror. Finn Askevold, "Measuring Body Image: Preliminary Report on a New Method," *Psychotherapy and Psychosomatics* 26 (1975): 71–77, on 73, 75. With permission from Karger. Copyright © 1964 Karger Publishers, Basel, Switzerland.

image-marking procedure developed by Norwegian psychiatrist Finn Aske-vold in a paper from 1975.[98] Askevold wanted to develop a "simple and inexpensive method for measuring body image" with an eye to patterns of distortions among healthy and pathological groups.[99] He placed a subject facing a wall with a body-sized piece of paper attached to it. The subject was then prompted to imagine herself standing in front of a mirror. The experimenter stood behind the subject and touched her at various points of her body; the subject then marked those body points on the paper in front of her, as she thought they would have appeared in the mirror (fig. 7.2, *left*). The points were then connected to form a coherent shape of the torso. Once this was complete, the experimenter drew the real silhouette of the subject over the subject's drawing for comparison (fig. 7.2, *right*). It was found that the shape drawn by an anorexic patient was larger than the one drawn by the experimenter; anorexics saw themselves as larger than they really were.[100]

The Askevold study visualizes very clearly the dual function of the mirror: what the patient saw and what the mirror reflected (though in this case, the experimenter-drawn figure assumed the reality function of the mirror). Again, this dual function was exploited for the diagnosis of pathology, which depended on the comparison and mismatch between the distorted inner body image (the solid line) and the real shape of the body (the dotted line).

Mirror Therapy

The ambiguous relationship between mirror and body image opened up the possibility for therapy.[101] For some patients, the mirror was an integral part of the disease. They felt compelled to check their appearance constantly in the mirror, thus getting trapped by the image. The great majority, in contrast, were horrified by mirrors and tried to avoid looking at their own reflections. The mirror image confronted them with their greatest fears. Although the first encounter with the mirror image might have dredged up long-standing anxieties and negative feelings, the mirror image also held the key to working on the body image, which, the researchers came to think, was malleable.

The first step in the mirror treatment was thus to help the patient overcome the initial emotional distress caused by the mirror.[102] This could be done in various ways. Most practitioners used relaxation of some form or another. Clinical psychologist Silja Vocks and colleagues at the University of Bochum in Germany, for instance, used relaxation *before* mirror confrontation: subjects sat in an armchair listening to music for five minutes and then stood with their backs to the mirror for another five minutes before starting the exposure. Others used relaxation *during* the exposure. For example, Thomas Cash and Jill Grant encouraged patients to use relaxation techniques (that they had learned previously) to manage their emotions as they progressively engaged with the areas of their bodies that gave them most distress.[103] To facilitate this engagement further, patients were first encouraged to imagine their bodies in their minds, to "progressively 'picture' disaffected body areas, from least to most disliked." Only after they had gained the ability to see themselves using their inner eye "with reasonable control of discomfort" were patients allowed to move on to the next step: mirror exposure.[104]

Other therapies required a more comprehensive cataloging and analysis of the emotional response, including that of the German psychologists Silja Vocks and Tanja Legenbauer, with whom I opened the chapter. Their protocols recommended pre-exercises that is, "imagination exercises." The patient was encouraged to "concentrate on [the] body and for 2 to 3 minutes collect those thoughts that had entered the mind during the concentration on the body."[105] They then wrote these thoughts down and discussed them with the therapist. Together they decided whether the notes reflected their thoughts or emotions. After patients had learned to differentiate between the two, the therapist tried to establish with them "that positive and negative statements trigger different sensations in the body."[106] To process this, patients were asked to "address one of the written-down statements and focus on them. The eyes should be kept closed. The patients should pay attention to how they feel when they focus on the statement."[107] Then, finally, patients were to pick a negative thought and explore what emotion was associated with it. They were encouraged to do the same with a positive emotion. Ultimately, "the goal is to raise the patients' awareness that a negative statement can cause a negative bodily sensation, a positive statement on the other hand is more likely associated with a neutral or positive feeling."[108]

With these insights established, the patient was ready to begin the actual mirror treatment. The goal here was for the patient to learn how to account for her body objectively. The patient was encouraged to describe herself from head to toe, according to what she saw in the mirror. The mirror was supposed to be "sufficiently large" so that the patient could see her entire body; ideally, it would be a "mirror with two wings so that the patient can look at her body more easily from different perspectives."[109] Often, she navigated the mirror image moving from the least disliked to the most disliked body part, in order to maintain a check on her emotional reaction.[110] In all this, the patient was encouraged to be as "objective" and neutral as possible. Some manuals recommend that the patients should imagine they were describing their bodies to a painter, to a blind person, to a person on the phone, or even to an "alien," in the attempt to reduce the emotions connected to the experience.[111] As Vocks and Legenbauer pointed out, "it is important in the exposure task to interrupt the patient each time when he makes negative statements." The point was to juxtapose the neutral description of the body to the emotionally freighted one, both of which the mirror seemed able to elicit, and to use the former to correct the latter. Slowly, assisted by the therapist, the patient could come to see a different image of herself in the mirror, one that was closer to the "real" body. The hope was that the external image would finally win out and overwrite the internal body image.[112]

In some approaches, the correction of the "irrational belief" was not the ultimate goal; the patient should learn to assess her body not simply in neutral terms, but in a positive way.[113] The patient was encouraged to draw on the mirror. One patient had produced the following description before treatment:

I have black, thin hair that hangs down straight and frames my moon face. My nose is rather crooked and my eyes are quite small. Like piglet's eyes. I have full lips and a dimple on the left cheek. My neck is rather sinewy and long, my shoulders broad and angular. My breasts (bra size A) are too small and my stomach is flabby and looks four months pregnant. My legs are too short for my upper body and are moreover much too fat. They look like elephant legs [*Stampfer*].[114]

To help the patient develop an alternative, more "positive," way of seeing herself, another group member—often this kind of treatment was conducted in a group—gave a positive description of the patient's body. This was supposed to "draw the patient's attention to positive aspects that he normally doesn't perceive."[115] In other words, the idea here was to attach a different, positive emotion to the negatively evaluated (thought-about) body part—ironically reproducing stereotypical markers of womanhood. Here is what this looked like:

> Head: Beautiful black shiny hair, aristocratic nose, kissing mouth, wonderful green eyes with long black eyelashes; trunk: muscular upper arms, long slim fingers, beautiful finger nails, beautiful cleavage, gorgeous waist and cute belly with a slight curve and a mole; hips/buttocks: tight buttocks, feminine curve; legs: not too short for upper body, proportions are good. Beautiful form, muscular but feminine; overall image: attractive woman with a lot of charisma and individual type.[116]

In the final stage the patient was to give this kind of positive description herself using the mirror. This part was also done as a homework in order to "stabilize" this positive experience of her body.[117]

Conclusion

In the mirror therapy, we see the patients struggle with their own subjective experience of their body. The "objective" or "neutral" image against which the patient's body image is measured gains its value, not because it offers some entirely detached inhuman perspective, but rather because it offers a counterpoint to this experience. In fact, it is telling that the patient is not alone, but is rather aided by a therapist; the patient is encouraged to see his or her body as the therapist sees it. That is, we are dealing not with what we might imagine to be a purely objective vision, but rather with an intersubjective one, revealing to ourselves our bodies as they are perceived by others. That was—and is—the special power of mirrors.

In exchanging one point of view for another, the countervailing image was not free from social distortions; it prioritized some shared values over others. While pushing back against the slimness ideal, the patient in Vocks and

Legenbauer's mirror therapy was encouraged to become complicit with other forms of objectification. This explains the uncomfortable sexualization that persisted in so many of these encounters: the "kissing mouth" and the "feminine curve." We might want to bring to mind here the feminist criticism of Bruch. The patient was asked to find ways to conform to certain social norms that should be the real object of our transformative efforts.

Whatever concerns we might have with respect to mirror therapy, however, we should not fail to appreciate its strengths. The body image has long been a complex and contested concept. Even those researchers who relied on it, including Bruch, found it difficult to say exactly what it was and how it worked. Mirror therapy, however, offered a means to see it, measure it, and even change it. It made the body image real in a way it had not been before. For whatever the true status of the "objective" and "neutral" reflection, the mirror also served as a space where our internal and otherwise inaccessible experience could be offered up to study and intervention. While Lacan saw the mirror image as an object of misrecognition, for the body-image researchers it was this and something more. Indeed, the diagnostic and therapeutic power of the mirror arose from its ambiguous status. Our reflections are vehicles for our innermost desires and anxieties, but they are also a space where those desires and anxieties can be shared, a space we don't need to examine alone. At its core, that is what mirror therapy boils down to: two people looking together at the same reflection.

If the mirror image could render available to intersubjective study and even manipulation something internal like a body image, a brain structure, then perhaps that brain structure might itself be a type of mirror. Though working in an entirely different field, and almost certainly without knowledge of each other, while researchers were developing mirror anorexia therapy, another group of researchers explored precisely this insight in the discovery and analysis of what would soon become known as "mirror neurons."

8

Imperfect Reflections

MIRROR NEURONS, EMOTION, AND COGNITION

THERE IS NO MIRROR this time; at least not one that is visible. It is 1995, and we are in a lab at the Istituto di Fisiologia Umana at the University of Parma.[1] A macaque is strapped into a chair. Around the monkey some scientists are preparing an experiment, while others are examining the results of a PET scan on computer screens across the room. The machines are monitoring a group of cells located in area F5 of the rostral part of the inferior premotor cortex (area 6), an area associated with motor control. In a series of tests, one of the scientists places a raisin on a small tray in front of the macaque, and then reaches out to grasp it. The monkey is then encouraged to grasp the raisin too. What fascinates the scientists is that the same brain cells are firing in both situations: when the macaque observes the raisin being grasped, and when the macaque grasps the raisin itself. The following year, the neurophysiologist Giacomo Rizzolatti, the lead scientist at Parma, christened these cells "mirror neurons" (fig. 8.1).[2]

Rizzolatti and colleagues had latched onto the term "mirror" because the neurons they were studying responded to two "mirrored" actions: an action that was merely observed, and thus virtual for the subject, and the subject's own physical movement. But if this was mirroring, it was a peculiar sort. First, the observed and the executed act were not simultaneous. Indeed, as we will see, the temporal separation of the two was a necessary condition of the discovery of and research on mirror neurons. This separation was achieved by a process of action inhibition, either externally through the experimental setup or internally through neurological processes. Second, the executed and observed actions were ordered by the self-other distinction, which resulted in an

FIGURE 8.1. Experimenter and monkey grasping a raisin, neural activity during both behaviors: "The neuron discharges during grasping observation, ceases to fire when the food is moved and discharges again when the monkey grasps it." Giacomo Rizzolatti et al., "Premotor Cortex and the Recognition of Motor Actions," *Cognitive Brain Research* 3 (1996): 131–41, on 133. Reprinted with permission from Elsevier.

irreducible "incongruence" between the two. When we looked into the neurological mirror, we didn't see ourselves, but someone else.

This was a reversal of the stance of researchers during the first blooming of the mirror test, when they sought for a secure ground to determine whether we saw in the mirror ourselves or someone else. In foregrounding the way we identified with another, distinct individual, Rizzolatti's work resonated rather with that of the mirror researchers I discussed in the second part of the book. The identification between the two actions performed by the mirror neurons involved a misrecognition. The similarities help explain why the mirror neuron experiments, despite the divergences with the other tests I have discussed, came to be intertwined with a number of familiar elements from across the history of the mirror-test tradition, including human-animal difference, imitation, the origin of language, and the relationship between the emotional and the cognitive. So too, mirror neurons were related both in the circumstances of their discovery and in their application to cerebral body maps, which are best seen as a somatic counterpart of the body images I discussed in the last chapter.

The appearance of neuroscience in our story should not be that surprising. As we have seen, at multiple points within the mirror-test tradition, psychologists felt compelled to draw on contemporary understandings of the human brain. Preyer turned to association neurology, seeing mirror recognition as an extension of the process by which the child slowly gained mastery of its own movements, and thus began to distinguish itself from the world. Even Lacan, in his mirror work, referred to the process of myelinization and appealed to the homunculus. For these researchers, mirror behavior was merely the external manifestation of cerebral changes. It is perhaps to be expected that the language and questions of the mirror-test tradition should find their way into some forms of neuroscience.

In this chapter, I work out these parallels with respect to mirror neurons. Through a close analysis of the experiments that led to their discovery, I show how, despite appearances, the structure of misidentification, understood as both a temporal disjuncture and an incongruence between the original and mirrored action, was constitutive of mirror neuron research. I then show how this structure worked when mirror neurons were used to understand empathy, debates about human-animal difference, and the origin of language. Finally, by examining the work of the neurologist V. S. Ramachandran, I show how mirror neurons could function as an intermediary between neurological and psychological worlds, by following how they shaped his reading of none other than Sigmund Freud.

The Emergence of Mirror Neurons
as a Research Program

Rizzolatti and his group had not set out to discover neuronal mirroring. Rather their research project emerged out of a prior interest in body maps. Over the course of the 1980s—Rizzolatti had been a professor at the Institute of Human Physiology since 1975 and slowly built up a team of PhD students and postdoctoral researchers, some of whom would later take up positions across Europe[3]—the Parma lab had been trying to prove that a part of the premotor cortex was responsible for complex "actions." The scientists distinguished "actions" from simple "movements," describing the former in a 1988 paper "as a sequence of movements which, when executed, allows one to reach his goal."[4] In focusing his argument on such actions, Rizzolatti and colleagues were working against a tradition of brain mapping that stretched back to the Canadian neurosurgeon

Wilder Penfield. In the 1930s, 1940s, and 1950s Penfield and his colleagues at the Montreal Neurological Institute (MNI) had mapped out the primary motor and somatosensory cortex, that is, the parts of the brain responsible for processing motor and tactile information—a neurological parallel to the "body images" I discussed in the previous chapter. The maps were configured pictorially, as so-called homunculi spread out on the surface of the brain; these images have since become iconic (fig. 8.2, *top*).[5] After Penfield similar maps were constructed for other animals including the so-called simiusculus of Clinton Woolsey at the University of Wisconsin (see fig. 8.2, *bottom*).[6]

A key characteristic of Penfield's maps was that they came in pairs, with a sharp spatial distinction between the sensory and motor cortex, each of which had its own somatotopy. The motor map indicated which part of the cortex controlled the movement of which body parts (such as the flexion of the biceps muscle). In this model, complex actions, which required the combination of multiple different movements (e.g., flexion of the biceps *and* rotation of the shoulder), could be explained only by appealing to a higher-order integrating system located between the sensory and motor cortices, regions that were sometimes called "associative areas."[7] These areas integrated information from the sensory cortex and transmitted it to several parts of the motor cortex. In this view then, the motor cortex was considered "peripheral and almost exclusively executive."[8]

In a set of papers published in 1988, Rizzolatti and his colleagues set out to demonstrate that such complex actions could be coded without reference to an association system. A prime candidate was anatomic area 6, located just anterior of Penfield's "primary" motor cortex.[9] To test this hypothesis, they conducted the following experiment. Three macaque monkeys, "selected for their docility," were trained to sit on a chair with their head fixed looking forward.[10] They were surrounded by a plexiglass perimeter at arm's length. The plexiglass had nine holes arranged in three rows into which the experimenters could place pieces of food. The monkey was then prompted to perform what was for Rizzolatti a complex action rather than simply movement: in this case the monkey extended its arm to reach for the food. Through a range of variations on the test—changing the location of the food, and holding it with different experimental apparatuses—researchers were able to relate a set of actions to the activity of particular neurons.[11] Based on these data, they mapped out the function of cortical brain area 6, developing maps such as the one shown in figure 8.3. The maps showed that the area had its own somatotopy, that is, it was distinct from the traditional motor homunculus, and, most importantly, it was a map of complex actions.

MOTOR SIMIUSCULI

FIGURE 8.2. *Top*: Wilder Penfield's motor and sensory homunculi. Wilder Penfield and Theodore Rasmussen, *The Cerebral Cortex of Man: A Clinical Study of Localization of Function* (New York: Macmillan, 1950), 44. Courtesy of the Osler Library of the History of Medicine, McGill University. *Bottom*: Clinton Woolsey's motor simiusculi. C. N. Woolsey et al., "Patterns of Localization in Precentral and 'Supplementary' Motor Areas and Their Relation to the Concept of a Premotor Area," *Research Publications—Association for Research in Nervous and Mental Disease* 30 (1952): 238–64, on 252.

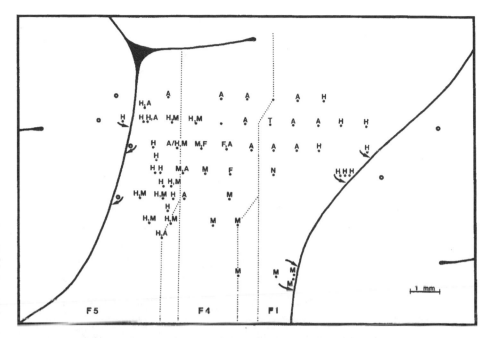

FIGURE 8.3. Brain map of areas F1, F4, and F5. The letters indicated parts of the body: A = arm; F = face; H = hand; M = mouth; N = neck; T = upper trunk. M. Gentilucci et al., "Functional Organization of Inferior Area 6 in the Macaque Monkey: I. Somatotopy and the Control of Proximal Movements," *Experimental Brain Research* 71 (1988): 475–90, on 478. Reprinted by permission from Springer Nature.

As the experimental setup suggested, the increased complexity of the motor cortex needed to understand "actions" rather than "movements" started to chip away at the sensory-motor divide. The tested actions were no longer simply motor but included the processing of a changing set of sensory stimuli as well—the visual recognition of the stimulus in space and the proprioceptive experience of grasping.[12] A complex action seemed to have a much more robust relationship to sensory stimuli than a simple movement. In brushing away a mosquito, for example, the action was not only initiated by the sensation but was also strongly determined by the bodily location of the sensory stimulus.

The combination of sensory and motor aspects in these actions had its corollary at the neuronal level. The team found a group of neurons in F4 that was "strongly responsive to tactile stimuli" as well as "related to proximal and facial movements"; forty-eight neurons in total fired both during movement and during tactile and visual stimuli.[13] Moreover the stimuli and actions were closely related: the neurons that controlled proximal movement (that is, movement within the space close to the body) were "triggered by stimuli presented in the animal's peripersonal space. . . . Thus, an object presented in a particular spatial position activates the neurons controlling the motor act 'reach' and, if the motivation is sufficient, these neurons will bring the arm in the space position where the stimulus is located."[14] The researchers concluded that, for F4 neurons, "their input-output relationship is very complex."[15] The difference between the primary motor cortex and area 6 was thus "quite clear."[16]

Inhibition and the Sensory-Motor Distinction

In the initial 1988 paper, Rizzolatti and his team did not examine the relationship between the sensory and motor responses of the neurons rigorously. The goal was simply to prove that a part of the premotor cortex was functionally complex. Nevertheless, the work marked an important and consequential shift from the traditional mapping project. No longer did the sensory-motor distinction, as in the Penfield model, correspond to anatomical separation. Now sensory and motor properties were transposed into a single neuron. The distinction between them would have to be figured in new, nonspatial ways.

In 1990 the team gathered further evidence for the complexity of area 6 neurons by turning their attention to F5 (which, like F4, belonged to the inferior part of area 6).[17] As before, these neurons seemed to be closely related to complex actions, in this case holding, grasping, or reaching for objects (which explains their name, reaching-grasping neurons). The crucial innovation of the 1990 paper was the introduction of time as a variable. The researchers distinguished two distinct stages of the neuronal activity, one sensory and one motor, delimited by three moments: the initial awareness of the sensory stimulus (marked by a saccade, which is an eye movement), the beginning of the movement, and the end of the movement. In the resulting visualizations of the results, the researchers marked these by a triangle, a dotted vertical line, and a heavy vertical line respectively (fig. 8.4).

Figure 8.4 represented the response of one type of reaching-grasping neuron. As it showed, the researchers noted a large change in neuronal activity—in this case inhibitory modulation (white space rather than dashes or dots)—after the presentation of the stimulus. In fact, this inhibitory modulation was present both in the premovement phase (from the presentation of the stimulus to the beginning of the movement, triangle to dotted line) and during the movement (dotted line to heavy line). The conclusion was that for this type of neuron (which constituted about half of those tested in the area F5), there was

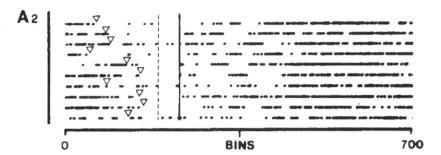

A 2

0   BINS   700

FIGURE 8.4. The triangles signify the saccadic eye movement triggered by stimulus presentation; the dotted lines designate the beginning of movement; and the solid lines designate the end of the movement. Ten individual trials were shown. Modified from G. Rizzolatti et al., "Neurons Related to Reaching-Grasping Arm Movements in the Rostral Part of Area 6 (Area 6aβ)," *Experimental Brain Research* 82 (1990): 337–50, on 342. Reprinted by permission from Springer Nature.

a clear continuity between premovement and movement activity, between the neuron's activity in response to the sensory stimulus and that connected to the movement itself.[18]

The need to analyze the relationship between neuronal activity in the premovement and movement phases explains the appearance of the first real mirror. In their 1992 paper "Understanding Motor Events: A Neurophysiological Study," Di Pellegrino, Rizzolatti, and colleagues continued their attempt to tease apart the multimodal properties of neurons in area F5, repeating their experiment with a new setup. Now food was placed under a geometric object in a box. The front door of the box consisted of a one-way mirror. When the monkey pressed a switch, the inside of the box was lit so that the mirror became transparent, presenting the monkey with the sensory stimulus. Provided he continued to press the switch, after a delay of 1.2 to 1.5 seconds, the door would open, and the monkey could reach for the object and thus access the food. The one-way mirror experimentally produced the three dividing lines that had been described in the earlier paper: the presentation of the stimulus (mirror door became transparent), the beginning of the movement (mirror door opened), and the end of the movement (grasping the object). But unlike the relatively messy and varying timings of the 1988 experiment, where the triangles didn't line up and there was no consistent gap between the presentation of the stimulus and the start of the movement, here it could be more closely controlled: the two phases, "seeing without moving" and "moving," were clearly separated in the experiment.

In the elaboration and comparison of these two phases we can already see the basis for a type of mirroring: neuronal activity in response to visual stimuli that was similar to activity prompting motor actions. For our purposes what is crucial is the temporal distinction between the two, which was a condition of this analysis. Because the sensory-motor distinction had been displaced into a single neuron—there were no longer purely sensory and purely motor neurons that could be tested separately and, at least in principle, simultaneously—neuronal activity had to be tested under two mutually exclusive conditions, one when the animal's own movement was inhibited (glass door shut), the other when it was not (glass door open). The action inhibition provided by the mirror box allowed the temporal separation between the presentation of the sensory stimulus and the motor response.

Incongruence and the Distinction between Self and Other

Although we see here a similarity in the nervous activity related to both stimulus and response, it doesn't make sense to say that one mirrored the other. In their attempt to describe actions rather than movements, the Parma researchers had increased the complexity of the motor responses. The sensory stimuli remained, however, simple: an object placed in a particular location in the monkey's visual field. Stimuli and response thus differed radically. The emergence of mirroring as an explicit theme required a chance event.

In running the experiment, the researchers periodically had to enter the scene to pick up the food or place it inside the mirror box. During this process, which was officially outside the limits of the experiment, the monkeys remained in place and the recording equipment on. The electrodes thus caught the activity of the monkey's F5 neurons in response to the researchers' actions. As the 1992 paper noted, "in the absence of any overt movement of the monkey," the behaving scientist activated a "relatively large proportion of F5 neurons," which had been active when the monkey performed the same movements.[19] In other words, the researchers had inadvertently substituted a new stimulus, one that was much more complicated than the one provided in the experiment. The observed actions (researcher reaching for the object) were now similar to the monkey's (monkey reaching for the object); they *mirrored* each other. Following this discovery, the researchers began "performing a series of motor actions in front of the animal."[20]

The increased similarity between observed and executed action made the "congruence" between them a central concept in the research. Already in

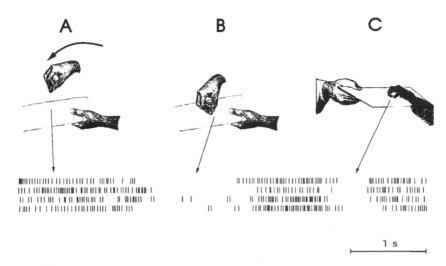

FIGURE 8.5. Example of a highly congruent mirror neuron. Giacomo Rizzolatti et al., "Premotor Cortex and the Recognition of Motor Actions," *Cognitive Brain Research* 3 (1996): 131–41, on 135. Reprinted with permission from Elsevier.

the 1995 experiment, Rizzolatti and his colleagues distinguished between two types of mirror neurons based on their congruence.[21] In the first class, the *broadly congruent* (the most common type, accounting for over 60 percent of mirror neurons), neurons could be activated by a varied range of actions, for example reaching, grasping, and rotation of the hand; in the researchers' words, "there was a link, but not identity, between the effective observed and executed action."[22] In the second class, the *strictly congruent* neurons, the observed movement, and the executed movement correlating with the neuronal activity "corresponded both in terms of general action (e.g. grasping) and in terms of the way in which the action was executed (e.g. precision grip)."[23]

This led to the experiment described at the beginning of this chapter. The neuron was tested under three conditions (fig. 8.5): in A the experimenter placed a raisin on a tray; in B he grasped the raisin; and in C the monkey performed the same grasping action. Actions B and C were "congruent," and in both cases the neuron showed a "response inhibition of the spontaneous discharge."[24] The action A was not, however, sufficiently similar to C, and in that case the neuron showed no change in activity. For "strictly congruent" neurons the observed actions had to reach a threshold similarity to the executed action in order for the neuron to fire.

Yet as this example showed, even with strictly congruent mirror neurons, the differences could never be reduced entirely. After all, the observed action was performed by a human and the executed action by a monkey. And because the monkey's movement was inhibited during the observation phase, the mirroring was necessarily structured by the self-other distinction.[25]

Mirror Neurons and the Social

Like many other neuroscientific paradigms in the contemporary period, mirror neuron research seeded a complex institutional ecology. From its core at the neurophysiology lab in Parma, it reached out across a transnational network of research institutes and working groups. As the institutional and scientific prestige of mirror neuron research grew, the Parma team entered into collaboration with other researchers: on the question of language Rizzolatti worked with the computer scientist Michael Arbib at the University of Southern California; for arguments about theory of mind, they collaborated with analytic philosophers such as Alvin Goldman at Rutgers University in New Jersey; for empathy, the Parma team drew on the imaging expertise of scientists including the Italian expatriate Marco Iacoboni at UCLA and the neuroscientist Bruno Wicker at Aix-Marseille University. These collaborations were supported by international grants such as by the Human Frontier Science Program (HFSP). Only recently has mirror neuron work been undertaken relatively independently of the Parma group, with research into autism and psychopathy (on which more below).

It was research on mirror neurons and empathy (a "mirroring" of other people's emotions) that garnered the greatest public attention, and contributed the most to the "hype" around mirror neurons.[26] The most famous paper was published in 2003 by Marseille-based neuroscientist Bruno Wicker in a collaboration with Rizzolatti, Vittorio Gallese, and Christian Keysers at Groningen, Netherlands (the latter were former graduate students and postdocs of Rizzolatti's): "Both of Us Disgusted in *My* Insula."[27] They ran separate fMRI trials on human subjects: one with subjects observing facial expressions of disgust and the other with subjects actually experiencing disgust through the exposure to "stinking balls." The researchers found that some regions of the anterior insula were activated in both cases. Employing the mirroring principle of the so-called classical visuomotor mirror neurons in F5, the findings suggested that the understanding of emotions took place by "matching felt and observed emotions" (see fig. 8.6 and plate 2).[28]

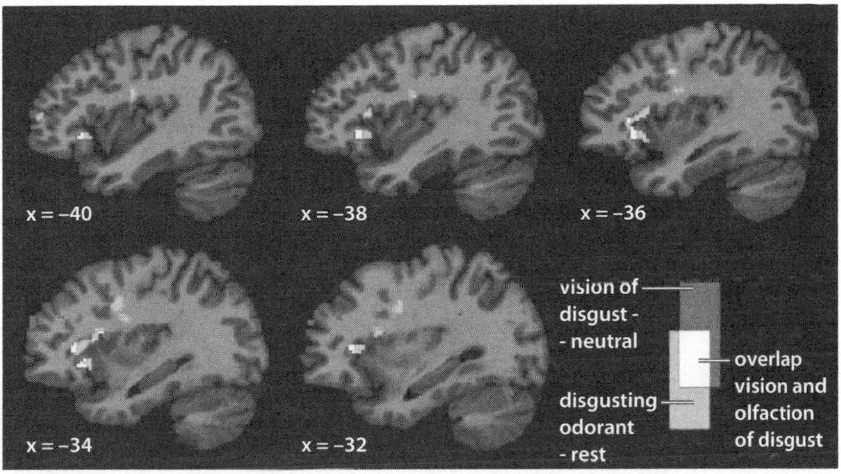

FIGURE 8.6. Disgust and mirror neurons: sagittal sections through the averaged left hemisphere of the fourteen participating subjects. The white patches indicate the overlap of the vision and olfaction of disgust, which are located in the insula. B. Wicker et al., "Both of Us Disgusted in My Insula: The Common Neural Basis of Seeing and Feeling Disgust," *Neuron* 40 (2003): 655–64, on 660. Reprinted with permission from Elsevier.

Wicker and his colleagues focused on the immediacy of empathy. In mirroring emotions, humans had a "hot," noncognitive understanding of others. The emotional and noncognitive nature of the process made it a likely candidate for the "evolutionary [*sic*] oldest form of emotion understanding."[29] In Wicker's words, it was a "'primitive' mechanism," shared by monkeys and humans to protect them from food poisoning.[30] Wicker's argument was part of a broader shift in neuroscience, in which researchers asserted the priority of direct and automatic processing of information.[31] Take for instance the 1998 paper by Vittorio Gallese and Alvin Goldman, a philosopher at Rutgers University, who sought to explain how "mind reading" (in the form of detecting certain mental states in others) was possible in humans. Gallese and Goldman favored "simulation theory" (ST) over the competing "theory theory" (TT).[32] According to theory theory, "mental states are represented as inferred posits of a naive theory."[33] Through a process of logical inference from a person's actions and appearance, the mental processes behind those acts could be understood and future behavior predicted. Here inference and cognition took center stage. By contrast, simulation theory suggested that we understood the actions of others through direct internal "representation" of the other's mental state, with no theorizing involved. Humans read other minds by "putting themselves in the other's shoes," feeling what they do simply by observing their facial and bodily demeanor.[34] It is easy to see how the mirror neuron system supported this latter model, because it suggested how observing an action could be neurologically similar to executing it. By extension, when we see the expression of an emotion we "feel" it too. Gallese was explicit in a paper published in 2001. To empathize, in his account, was essentially to understand: "My proposal is that also sensations, pains and emotions displayed by others can be *empathized, and therefore understood*, through a mirror matching mechanism."[35]

The pervasiveness of such claims, and the way they prioritized emotional over cognitive responses, has led scholars such as Ruth Leys and Allan Young to criticize mirror neuron researchers.[36] Mirror neurons, in this argument, offered a mechanism for understanding that did "not involve 'propositional attitudes' or beliefs about the emotional objects in our world. Rather, they are rapid, phylogenetically old, automatic responses of the organism."[37] In mirroring other people, we don't think about their situation but rather feel what it is like to be them. In this way, Leys has argued, the elevation of mirror neurons to a sort of human quintessence has tended to prioritize excessively the emotional and the immediate.[38]

Though the empathy experiments have attracted much public attention, they are actually an outlier in the paradigm. They are distinctive because of the way they elide (but never completely negate) the two forms of mediation central to the earlier mirror neuron studies. First, by shifting attention from "actions" to "emotions" the researchers sidelined the question of inhibition. Though the "real" emotion could be elicited (the stinking balls) alone, the internal nature of emotions made its presence more difficult to rule out empirically during observation. What did it mean to talk about "inhibition" in this context? Indeed, as we have seen, it became an assumption of empathy research that there was no inhibition; we feel the emotion when we see it. Second, the simultaneity of the observed and executed emotions also sidelined questions of their congruence. How could we measure the similarity between the emotional states in two different people?

Nevertheless, since the empathy work depended on earlier studies of mirror neurons these differences could not be effaced entirely. The importance of such differences is clearest in the researchers' attempts to understand disturbances of empathy: most importantly psychopathy and autism.[39] Psychopathy has been the object of a small but vibrant strand of mirror neuron research, most prominently in two groups: a collaboration between the University of Montreal and Harvard Medical School coordinated by the neurologist Alvaro

Pascual-Leone, and a research group at the University of Groningen in Holland led by Christian Keysers.[40] Mirror neurons helped explain the dual nature of psychopathy: psychopaths seemed both acutely aware of the emotions of others, which allowed them to be so charming, and indifferent to them, which accounted for their propensity for violence.[41] Relying on fiction to tell his truth, Keysers used Hannibal Lecter from *The Silence of the Lambs* to describe the condition: "Despite the varnish of sophistication . . . Lecter is capable of horrible crimes." What makes psychopathy so disturbing to us is not simply the discovery that psychopathic individuals lack empathy; rather, it is the insight that they "combine a talent for manipulation with a lack of remorse."[42]

Keysers's graduate student Harma Meffert and their colleagues demonstrated this dual character of psychopathy and its relationship to cerebral function in an fMRI study on eighteen convicted psychopathic offenders in Holland.[43] Dramatizing their study, Meffert and colleagues declared that they were dealing with "one of the largest group of psychopathic offenders ever scanned at 3 T,"[44] and had to move their subjects to the scanning facility "one by one, in bulletproof minivans."[45] Because no metal could be brought near the magnetic imaging scanner, the guards did not carry any firearms, but the patients "had wooden sticks sewn into their trousers and plastic handcuffs to keep them from running away or hurting anyone."[46]

In the study, the group sought to confirm the hypothesis that the psychopathic group would show less brain activity in mirror neuron areas—what they termed "vicarious activity"[47]—than a control group. That is, they wanted to show that psychopaths didn't empathize. In the "observation condition," the subjects were shown videos of human hands interacting with each other, under four conditions: love (with the hands caressing), pain (one hand hitting the other), social exclusion (one hand pushing away the other hand), and neutral videos (approaching hand touching the other and "getting a non-emotional response"[48]). In the "experience condition," the experiment got real: the subjects' hands were treated in the same way, being alternatingly caressed, hit or shaken, and pushed away by the experimenter. A third condition allowed the researchers to draw conclusions about what has been called the seductive-merciless character of psychopaths, what they referred to as the "empathy condition": here, subjects were explicitly instructed to empathize, to "feel with the receiving . . . or the approaching . . . hand" on the videos.[49] During all conditions, the brain activity of the subjects was recorded.

The study showed that the psychopathic group showed "reduced vicarious activity" among others in regions involved in experiencing emotions; the psychopaths did not "feel" the emotions of the people they observed. But this was not always the case. In the third "empathy condition" researchers found that "explicit instructions to empathize significantly reduced the group differences with regions associated with vicarious activations."[50] The mirror neuron system in psychopaths, which was normally "switched off," could be switched on.[51] This offered an explanation for the supposed ability of psychopaths to be charming as well as cruel, to be seductive as well as sadistic. It also offered new possibilities for treatment, as Keysers and his group pointed out: therapies could "harvest the patients' potential to normalize vicarious activations through deliberately focusing attention on empathizing with others."[52]

The Montreal-Harvard group led by Pascual-Leone cast this duality in slightly different terms. In their 2008 study, they recorded motor evoked potentials (MEP) induced by transcranial magnetic stimulation (TMS) in a group of eighteen male college students. Based on results from previous studies, they assumed that the observation of pain in others led to a decreased MEP amplitude in normal subjects, which was considered a measure of mirror neuron activity. All subjects also took a test to determine character traits: the Psychopathic Personality Inventory (PPI), which was divided along several axes including Machiavellian egocentricity, coldheartedness, and stress immunity. Somewhat surprisingly, the team found that individuals who scored highest on the coldheartedness scale displayed the largest modulation of cortical excitability. In their words, "increased levels of psychopathic personality traits in healthy individuals [were] clearly not associated with reduced activity within the MNS [mirror neuron system] for pain." Rather, the mirror neuron systems in those individuals showed an increased activity.[53]

To explain their results, the Montreal-Harvard group appealed to a distinction made by A. Avenanti and colleagues in 2005 between "sensory empathy" and "emotional, state or trait empathy."[54] "Sensory empathy" was the "strict ability to understand the affective, sensory or emotional state of another individual"; "trait empathy" made that information "available to the observer for an emotional/affective response." They suggested that this distinction— essentially one between recognizing and feeling emotion—was consistent with data from R. J. Blair who in 2005 had suggested that in psychopaths, "motor empathy" and "theory of mind" were unaffected whereas "emotional empathy" was impaired (here "motor empathy" roughly mapped onto "sensory empathy" in Avenanti's distinction). The high modulation of cortical excitability among the "cold-hearted" students suggested that they were sensorially empathetic, which enabled their "notorious manipulative nature" and "their ability to exploit weaknesses in others."[55] In contrast "trait empathy" seemed to be "maladaptive" in psychopaths.[56]

The two arguments relied on a similar set of structuring oppositions that we have followed throughout mirror neuron research: psychopaths lacked an emotional mirroring but were successful at non-emotional mirroring. Their pathology could be traced to their ability to distance themselves emotionally from the actions of others, to separate the knowledge of the observed act from the feelings they had themselves. The psychopaths could mirror the emotions of others, while inhibiting those emotions in themselves. The inhibition central to "sensory empathy" thus became key to understanding psychopathy.

While psychopathy research suggested the existence of a form of empathy and mirroring that was neither immediate, nor emotional, a parallel strand of research into autism suggested that these forms of difference—both inhibition and incongruence—were not merely pathological but were necessary for normal mirroring. Experimental evidence for a connection between mirror neurons and autism came from research performed by two groups in the middle of the first decade of the 2000s.[57] The first group, led by the neuroscientist V. S. Ramachandran and his colleagues at UC San Diego, has been one of the few who contributed in the debate about mirror neurons without sustained collaboration with the Parma center.[58] In 2005, Ramachandran used a new method involving "mu oscillations" to determine mirror neuron activity, comparing normal and autistic patients who opened and closed their own hands at a specific rate, while observing the same action on a video screen. The lack of "mu suppression" in the autistic patients suggested to Ramachandran a "possible dysfunction in the mirror neuron system."[59] In a popular science publication he presented this as the "broken mirror" hypothesis.[60]

The child and adolescent psychiatrist Justin Williams at the Department of Child Health at the University of Aberdeen and his colleagues (among them the evolutionary psychologist Andrew Whiten and David I. Perret, an expert in monkey neurophysiology at St. Andrews) published the second study providing experimental evidence for a link between mirror neurons and autism in 2006. Building on a protocol used by Marco Iacoboni and colleagues in 1999 to identify the neural substrate of imitation and mirror neuron function, they asked two (age-matched and IQ-matched) groups of subjects—one with autism spectrum disorder (ASD), one without—to lift their index or middle fingers according to three experimental conditions: (a) "animation," where they saw an index or middle finger raised up; (b) "symbolic," where either index or middle finger was marked with a cross; and (c) "spatial," where a cross

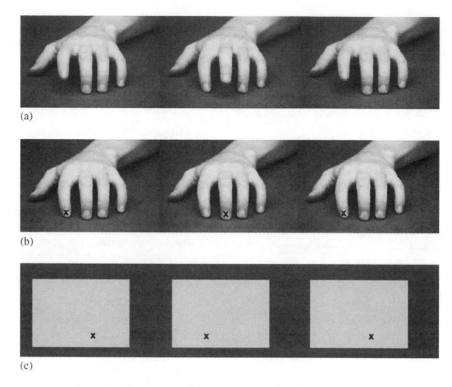

(a)

(b)

(c)

FIGURE 8.7. Justin H. G. Williams et al., "Neural Mechanisms of Imitation and 'Mirror Neuron' Functioning in Autistic Spectrum Disorder," *Neuropsychologia* 44 (2006): 610–21, on 612. Reprinted with permission from Elsevier.

was marked on a card (fig. 8.7).[61] The first condition served to test "imitation" whereas the second and third tested "execution," which acted as controls. In addition, in the "observation condition," the subjects were shown the three stimuli and asked not to move their hands at all.

During all these tasks the brain activity of the subjects was recorded in an fMRI scanner. In their results the researchers noted "robust evidence for differences between a control and an ASD group in the patterns of brain activation associated with imitation."[62] Most importantly, they found that the anterior parietal areas (that is, the mirror neuron areas) showed less activation in the autism group. But they also found areas of increased activity in the autism group during imitation, namely in the dorsal premotor cortex and parts of the dorsal prefrontal cortex (areas just outside the mirror neuron areas). The researchers interpreted this as the autism group "relying more on visuomotor learning" to perform imitative tasks.[63] They pointed to the various neural connections integral to proper mirror neuron functioning (and by extension, imitative behavior), which in autism were disturbed. A group of researchers around Riitta Hari at the Helsinki Institute of Technology noted a slightly different issue: individuals with Asperger's syndrome (a condition related to autism) manifested delayed activation of the mirror neuron system in the inferior frontal lobe during an imitation task.[64]

As we can see in these accounts, the authors discussed in considerable detail the neurophysiological changes seen in the disturbed mirror neuron system. But on the specifics of the failures (behavioral, cognitive, emotional, etc.), they were less clear. They did not specify how the symptoms of autism could be explained by a failure of mirroring. What did it mean to say that autism was caused by "broken mirrors"? For the most part, Ramachandran and Williams outsourced this part of the project, relying on other studies to show the breakdown of mirroring in autistic patients. Claiming that mirror neuron impairment helped explain the full range of symptoms observed in autism, including those involving imitation, language, theory of mind, and empathy, Ramachandran cited a range of works for each field.[65] Similarly, Williams referred back to one of his earlier articles on the topic, which contained a brief survey of autism research [66]

Williams's 2001 literature review showed the multifaceted relationship between autism and mirroring. He referred to studies showing how "people with autism do not readily imitate the actions of others," especially when those actions were "complex."[67] But while this problem fit with the idea that we have a breakdown of mirroring in autism, the authors also raised a different issue, referring to studies that described how autistic patients showed "inflexible and stereotyped behavior and language," such as "copying actions," "obsessive desire for sameness," and "more stereotyped mimicking, such as echolalia." For example, autistic patients might only repeat a question, instead of answering it. Williams's explanation for this excessive mirroring should hold our interest, because it recalls our previous discussion of the mirror neuron experiments. He argued that it might be a failure of inhibition.[68] The implication was that healthy mirroring required this inhibition, a delay between observed and executed action.

We see a reference to the second form of difference, incongruence, in a later paper by Ramachandran and his colleagues at the UCSD Center for Brain and Cognition. In an editorial for the journal *Medical Hypotheses* published in 2007, Ramachandran emphasized the ways in which mirror neurons allowed the subject to find commonality across different realms.[69] Ramachandran then extended the argument to help explain metaphor. Quoting Shakespeare's line "Juliet is the sun," Ramachandran claimed that we could understand it because we can "reveal the common denominator of radiance and warmth."[70] In autism, this ability fails; individuals with autism were often unable to understand metaphors and interpreted them literally.[71] As before, we see indeterminacy with respect to the failure of mirroring. Did the autistic patients lack functioning mirror neurons? Or was the problem that those mirror neurons were (in the language of the first paper) too congruent, and were activated only when the similarity between the two elements reached a certain threshold? As for the psychopathy research, the implication was that normality was marked by a certain tolerance for incongruity between observed and executed actions. The two didn't need to be exact replicas.

Autism thus seemed to be the inverse of psychopathy; the former promoted an excessive identification with the other, the latter made the separation too absolute. And again in the analysis, inhibition and incongruence were the guiding principles.[72] As the Canadian neuroscientist Shirley Fecteau and colleagues pointed out in a 2008 paper, whereas in psychopathy, emotional empathy was impaired with theory of mind and motor empathy intact, in autism, emotional empathy was spared, and theory of mind and motor empathy showed deficits.[73] The two conditions were mirror images of each other.[74] The analysis of autism suggested that normality might require a form of distancing and incongruence; the study of psychopathy suggested that this distancing should not be pushed too far. In both cases the diseases were understood through a refiguring of the two forms of mediation at the heart of the mirror neuron project.

Human-Animal Difference and the Origin of Language

The structure of mirror neuron research proved particularly valuable when it was deployed to address a question that as we have seen was central to the mirror-recognition-test tradition: evolution and human-animal difference. The history of mirror neuron research, as I have been telling it here, certainly seems to challenge the idea that mirror neurons set humans apart from animals. First, the function of mirror neurons turned on its head the very models of animal interaction that had previously been used to relegate nonhuman animals to a different realm. Mirror neurons blurred the division between self and other, such that it no longer made sense to ask whether the subject saw itself or another in a mirror. The answer was always both. Second, as we have seen, mirror neurons were first discovered in monkeys. The relationship becomes more complicated when we remember that the first mirroring took place between humans and animals, when the recording device picked up neuronal activity in the macaque brain as the scientists set up the experimental system.[75]

The interconnections between animals and humans were crucial as scientists shifted their research projects to study mirror neurons in the latter. The first experiment to show *avant la lettre* that mirror neurons might also be found in humans was presented in a 1995 paper, but the key paper appeared a year later.[76] The Parma group had teamed up with researchers at the University of Milan to conduct a PET (positron emission tomography) imaging study, measuring human brain activity under two conditions: grasping observation and object "prehension," that is, grasping.[77] The initial results were not encouraging. The brain images produced for observed and executed actions did not match. For example, while the inferotemporal cortex and the inferior frontal gyrus were activated by grasping observation in the experiment (see the second row of brains in fig. 8.8 and plate 4, *top*), no activation was found in the inferior frontal gyrus for grasping execution (or "object prehension," fig. 8.8 and plate 4, *bottom*).

The failure to match the brain activity under the two conditions was in part an effect of the recording equipment used. In humans, experimentation on the brain is tightly regulated; ethics protocols prevent researchers from simply opening up the skulls of human subjects to allow single cell recordings in the way they can with monkeys.[78] Instead the group resorted to the much less precise technique of PET imaging, which allowed the identification of activity in populations of cells in broadly defined brain areas. This imaging method was also much less sensitive. The reason why area 44 did not light up in the study, Rizzolatti argued, was that "the behavioral protocols . . . were not sufficiently demanding" to induce sufficient activity there.[79]

In light of this lack of evidence, scientists relied on homologies between monkeys and humans to argue for the existence of mirror neurons in humans. Much of the 1996 paper details such homologies in the inferior frontal gyrus, one of the areas they had found to be activated during human grasping observation in their study: human inferior area 6 was the homologue of monkey area F4, and human area 44 was the homologue of monkey area F5; for area 45, the situation was less straightforward.[80] Later we will see the significance of these homologies, especially of area 44, also known as Broca's area.

The homologies allowed the scientists to make the argument that, despite the results from the PET scans, the neurons activated in observation were also

Grasping observation vs object observation

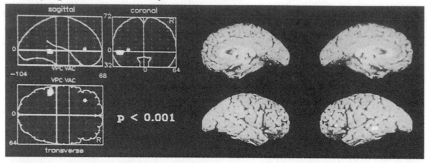

Object prehension vs object observation

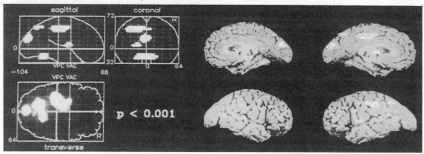

FIGURE 8.8. The inferotemporal cortex and the inferior frontal gyrus were activated by grasping observation (*top*: second row of brains); no activation was found in the inferior frontal gyrus for grasping execution (*bottom*: "object prehension"). G. Rizzolatti et al., "Localization of Grasp Representations in Humans by PET: 1. Observation versus Execution," *Experimental Brain Research* 111 (1996): 246–52, on 248. Reprinted by permission from Springer Nature.

active in the performance of an action. After noting that "no activation was found in the inferior frontal gyrus" during grasping movements, they pointed out that in the monkey, "the premotor cortical area that appears to be most crucially involved in grasping movement organization is area F5," the area of mirror neurons.[81] Referring to other evidence in humans (such as recent PET studies that had found increased blood flow in the area during the execution of movement sequences), they suggested that the area was, in fact, involved in the organization of complex motor tasks.[82]

Despite the closely entwined nature of human and animal research in the mirror neuron paradigm, mirror neurons have often been declared the key to what makes us human. The most prominent scientist to make this sort of argument is V. S. Ramachandran, who as we saw had played an important role in autism research. In 2007 Ramachandran discussed the so-called great leap (sometimes called the "big bang") of human evolution—the sudden rise in technological development and cultural expressions such as cave art, clothes, and new kinds of dwellings about forty thousand years ago.[83] He then posed the question why this happened only then, rather than 250,000 years ago, when the hominid brain had reached its present size. The common answer was that because of a sudden genetic shift, previously unconnected functional areas of the brain were suddenly able to work together. Ramachandran suggested a modification of this view. If there was a genetic change, it concerned the mirror neuron system, increasing its "sophistication" and therefore its "learnability."[84] In a 2009 TED talk, Ramachandran declared that they were the "neurons that shaped civilization."[85]

It is not immediately clear what Ramachandran meant by the "sophistication" of the mirror neuron system, but from the context we can suggest that he was referring to the human ability to mirror highly complex actions. These "sophisticated" mirror neurons, and the advanced imitation capacities they enabled, allowed developments such as "tool use, art, math and even aspects of language" to "spread very quickly through the population."[86] The emphasis here was not so much on invention but dissemination. Ramachandran admitted that the same or similar innovations might as well have occurred with "earlier hominins" like *Homo erectus* or Neanderthals, but, he suggested, because these hominids possessed a less powerful mirror neuron system, the innovations "quickly drop out of the 'meme pool.'"[87] In a parallel argument in 2007, Ramachandran extended the argument to link mirror neurons to self-awareness. The nature of the self, to Ramachandran, was essentially self-reflecting: a "sense of 'introspecting' on your own thoughts and feelings and of 'watching' yourself going about your business—*as if* you were looking at yourself from another person's vantage point." Self-awareness was mirroring "turned inward to look at your own self."[88]

In both cases—the great leap forward and self-awareness—the difference between humans and animals, and between *Homo sapiens* and other members of the same genus, seems to be located in the greater immediacy of the mirroring. Either mirroring was more fine-grained, able to imitate more fully the acts of others, or the self-other distinction was collapsed such that the observed and executed actions were performed by the same subject. Humans differed from animals not because they could mirror, but because *they could mirror better*.

The situation was, however, not so simple. The analysis of such higher-order mirroring was picked up by Marco Iacoboni at UCLA. In a long-standing collaboration with the Parma group, Iacoboni found evidence for a new type of mirror neuron that outperformed classical mirror neurons with respect to the imitating behavior they encoded. Iacoboni labelled them "super mirror neurons." The super mirror neurons worked as a "functional neuronal layer 'on top of' the classical mirror neurons, controlling and modulating their activity."[89] According to Iacoboni, however, and important for my purposes here, these "super mirror neurons" were distinguished not by the way that they reduced the difference between observed and executed actions but rather by the opposite.[90]

To study the mirror neurons, Iacoboni took advantage of a group of patients suffering from drug-resistant frontal lobe epilepsy. The last-resort treatment for these patients was surgery, and in order to locate the exact focus of the epileptic activity and tell the surgeon where to cut, electrodes were routinely implanted in various brain regions, so that brain activity could be measured over an extended period of time. The situation provided an unrivaled opportunity for studying this area of the brain with far greater precision than was usually allowed; Iacoboni could benefit from the "exquisite resolution of single cells" recorded from these patients while they completed various tasks.[91] In the observation conditions, subjects were asked to observe various grasping actions and facial gestures on a computer screen; in the execution conditions, they were cued to perform the same actions in response to a visually presented word.

Again the researchers identified a set of mirror neurons, which responded during both the execution and the observation of the same action (e.g., smile observation and smile execution). These however could be divided into two groups: one that had similar responses in both action execution and action observation, and another where the cells responded with "excitation during action-execution and inhibition during action-observation" (fig. 8.9 and plate 5).[92] It was the latter group that Iacoboni called "super mirror neurons," which, the researchers suggested, might play a role in controlling "unwanted imitation." As they wrote:

> Mirroring activity, by definition, generalizes across agency and matches executed actions performed by self with perceived action performed by others. Although this may facilitate imitative learning, it may also induce unwanted imitation. Thus, it seems necessary to implement neuronal mechanisms of control. The subset of mirror neurons responding with *opposite* patterns of excitation and inhibition during action-execution and action-observation seem ideally suited for this control function.[93]

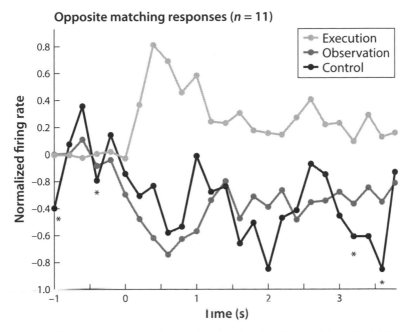

FIGURE 8.9. Super mirror neurons showing inverse mirroring. Reprinted from Roy Mukamel et al., "Single-Neuron Responses in Humans during Execution and Observation of Actions," *Current Biology* 20 (2010): 750–56, on 754. With kind permission from Elsevier.

Moreover, though here the argument was less clear, the ability to modulate imitation, to control and differentiate between the observed and executed action through a process of inhibition, seemed to be a condition of higher-order mirroring.

Though "super mirror neurons" seemed to be able to mirror better, Iacoboni explained that improvement by their ability to separate the observed and the executed act. But as we know, this was not a new element in the mix. In breaking the immediacy of the relationship between observed and executed act in super mirror neurons, Iacoboni was only maximizing a similar mediation at the heart of all mirror neurons: the *action inhibition* during observation that had been central to their discovery in the first place. That is, "super mirror neurons" were able to explain complex mirroring and thus to mark the difference between humans and animals because they were able, by drawing on a principle of inhibition, to exploit the mediacy that was inherent to the mirror neuron project.

It is in this context that some mirror neuron researchers homed in on a question that has dogged this book from the beginning: the origins of language. This was not a widely studied question in mirror neuron research; we are essentially dealing with a handful of papers written in the 1990s, of which a short speculative paper by Rizzolatti and the USC computer scientist Michael Arbib from 1998 is the most important.[94] But the arguments used are nonetheless instructive and can be seen as an extension of the earlier mirror tradition, which was concerned with the problem of identification as well as the analysis of mirroring. While the assertion of human-animal difference in the case of higher-order mirroring can be seen as the effect of the temporal splitting of observed and executed act through inhibition, in the case of language, that difference was due to the second form of mediation: a necessary tolerance for incongruence between the two.

As mentioned in my earlier discussion of the 1996 PET experiment, the Italians had argued for the existence of mirror neurons in humans by drawing on homologies with monkey brains. Significantly, F5 in monkeys, the area that contained mirror neurons, corresponded to area 44 in humans, also called Broca's area. Since Broca's area has long been regarded as the area of the brain related to motor speech, the homology seemed to suggest that mirror neurons might play a role in the emergence of language. To understand this connection, Rizzolatti took up the idea that there was a sharp neurological divide between human and animal communication. Vocalization in nonhuman primates was mediated by the cingulate cortex and, in particular, the diencephalic and brain stem structure, located deep within the brain, far from Broca's area, which was responsible for speech in humans. In addition, animal vocal calls seemed to serve a different form of communication, which aimed not at a particular individual, as did speech, but at a group at large. Finally, monkey vocalization was related to instinctive and emotional behavior, whereas human speech was not.[95] Rizzolatti used this argument to suggest that monkey vocal communication might not be the most proximate ancestor for human speech. Rather human linguistic ability might have evolved from animal gestures, gestures controlled by mirror neurons. The mirror neurons in motor area F5 might then be the "neural prerequisite for the development of inter-individual communication and finally of speech."[96] This was the significance of the supposed homology between F5 and Broca's area.

Rizzolatti argued that mirror neurons could explain primary communicative gestures, but to do so he had to rely on the incongruence between the mirrored actions. When confronted with an action "of particular interest," the mirror neuron system would "allow a brief prefix of the movement to be exhibited."[97] Thus in seeing ingestive processes, the monkey would perform "lipsmacks," or "tonguesmacks."[98] In similar ways, "a primitive vocabulary of meaningful sounds could start to develop."[99] Later experiments suggested a similar process in humans. For example, Gentilucci and colleagues presented participants with two 3-D objects, one large, one small, and asked them to open their mouth when seeing them. They found that lip aperture increased when the movement was toward a large object, and decreased when directed at a small object.[100] What is important for my argument here is the way in which differentiation or the lack of absolute congruence was central to the process: the mirrored "prefix" (lipsmacks) was only weakly congruent with the act observed (eating). Such differentiation allowed the connection yet difference between the signified object and the signifying sign, and, by extension, the development of a linguistic system where even this tenuous relationship of resemblance was given up. That is, the combination of multiple noncongruent mirrorings eventually produced a purely conventional relationship between signifier and signified. In Rizzolatti and Arbib's words, "sounds acquired a descriptive value."[101] Mirror neurons thus came to explain the development that for Tiedemann had marked humans' linguistic advance over other animals.

V. S. Ramachandran and the Mirror Box

The mirrors I have been discussing in this chapter have been mostly metaphorical, but in the early 2000s the mirror neuron tradition collided with a set of material research practices that might look much more familiar. The figure at the center of this convergence was someone whose name has already appeared in this chapter: V. S. Ramachandran. Ramachandran related the mirror neurons to his earlier work on "mirror box therapy" for patients suffering from phantom limb pain, which he first described in 1994.[102] Ramachandran argued that phantoms arose because of a misalignment between our physical body and our internal "dynamic body image," a cerebral structure that often survived after amputation (and neuroscientific analogue to Bruch's body image).[103] This body image was so powerful that it strongly resisted visual evidence that a limb was missing; patients still felt the absent arm or leg, and it could cause them significant discomfort.[104] The dominance of image over reality was not total; rather the characteristics of the phantom limb were structured by its relationship to the real body. One patient, D. S., considered his phantom limb to be paralyzed, because despite his attempts to move it, sending out a motor command from his motor brain map, he received no proprioceptive or visual evidence that it had changed position. As Ramachandran wrote, "Eventually, the brain learns that the arm does not move and a kind of 'learned paralysis'

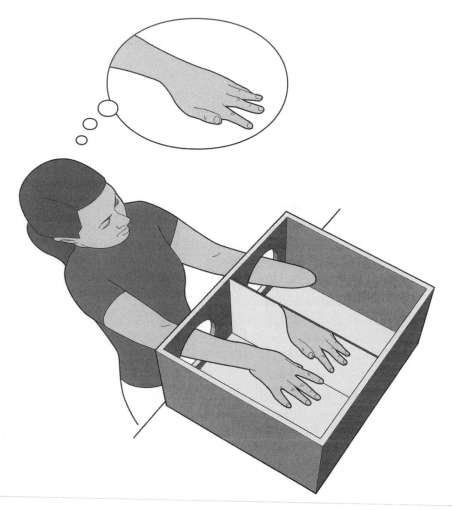

FIGURE 8.10. V. S. Ramachandran's mirror box. V. S. Ramachandran and Diane Rogers-Ramachandran, "It's All Done with Mirrors," *Scientific American Mind* 18, no. 4 (August/September 2007): 16–18, on 17.

is stamped onto the brain's circuitry."[105] Another patient, R. T., felt pain in his hand because without visual evidence he found it difficult to convince himself that his phantom hand could be unclenched.

Ramachandran's mirror box was designed to realign the body image with the physical body. It was constructed by placing a mirror vertically in the middle of a cardboard or wooden box.[106] For those patients with a phantom arm, they would place their healthy arm into one side, and the phantom limb into the other. As shown in fig. 8.10 and plate 6, the top and front side of the container were open, which allowed the patient to look into the box, but for the purposes of the treatment she had to cock her head to one side such that she could see the mirror from the side of the box with the healthy arm. Next, the patient was asked to move her healthy arm around until its mirror image "superimposed on the felt position of the phantom."[107] If working properly, when patients performed "mirror-symmetrical movements" they would see their phantom arm "resurrected."[108] As one patient exclaimed in surprise: "Mind-boggling. My arm is plugged in again; it's as if I am back in the past. . . . It no longer feels like it's lying lifeless in a sling."[109]

Though the chain of relationships is long and complex—the right healthy hand reproduced the movements of the phantom, the mirror produced an immaterial reflection of that movement, the patient took the immaterial mirror image to be the real physical left hand, and that supposedly physical left hand compared to the phantom—at its most basic level, the mirror created an immaterial double within the experiment that could be aligned with the body image in the brain.

Ramachandran appealed to Rizzolatti's "mirror neurons" to explain how the mirror box functioned: They controlled the alignment between mirror image and body image. For mirror neurons produced their own "virtual realities," a neurological simulation of observed action.[110] As Ramachandran and his collaborator Eric Altschuler stated, "mirror neurons necessarily involve interactions between multiple modalities—vision, motor commands, proprioception."[111] They transformed visual input into proprioceptive or tactile input. That is why, when patients saw their paralyzed phantom move, they could feel them move as well. Moreover, Ramachandran argued that mirror neurons could explain how vision was able to reshift the brain map, in order to make the immediate effects of the mirror box treatment longer lasting. Visual input could stimulate motor neurons on the body map that had been left "inhibited" or "dormant" after the amputation, recruiting them for new purposes.[112] That is, the existence of mirror neurons opened up a way to access neural material that had lost the possibility for direct sensory stimulation.

In his explanation, then, Ramachandran posited an intimate relationship between the body maps that lay behind the phantom limb phenomena and mirror neurons. The virtual reality produced by the mirror neurons, an imaginative copying of the action in the brain, resembled in this way the dynamic body image and indeed often occurred in precisely the same brain area. After all, the maps and body images responsible for phantom limbs were in their way their own mirror images (if sometimes distorted) of the moving body. The

, similarity explained why the "virtual realities" produced by mirror neurons could help mold the body image: the body image was a mirror image too. And at one further step, as for the body-image disturbance researchers we discussed in the previous chapter, the mirror became a way for Ramachandran to manipulate and thus experiment on the virtual images created by the mirror neurons. The mirror made the phantom present, malleable, open to study. It was a meeting point between the real and the virtual, the body and its phantom, the material and the immaterial, a point of conjuncture of different ontological spheres.

Conclusion

Ramachandran stands out as a maverick in contemporary neuroscience. He has built his reputation, including being named as one of *Time*'s one hundred most influential people in 2011, on his ability to pick up and run with trends and arguments merely pointed to and suggested by his colleagues—occasionally, some critics might suggest, beyond what the evidence allows.[113] But in highlighting ideas and possibilities left out by his more restrained peers, he also serves a valuable diagnostic role, bringing out the ways in which neuroscience engaged and often struggled with concepts and vocabulary from the mind sciences. One peculiarity of his writings is that they are haunted by someone at best ignored and at worst disparaged by most neuroscientists: Sigmund Freud. The father of psychoanalysis is mentioned with insistent regularity throughout Ramachandran's articles, books, and television appearances. We find reference to key Freudian concepts such as repression and the unconscious in his writings targeted at a popular audience, but also, and more surprisingly, Freud crops up continuously in Ramachandran's more narrowly scientific publications.[114]

Ramachandran's use of Freud is motivated by the insight that psychoanalysis could furnish contemporary neuroscience with a number of useful concepts, if only they could be translated into rigorous neurological language. He was insistent that Freud "was originally trained as a neurologist," and though for various reasons he had to depart from this path, Freud had "never lost sight of his initial goal of providing a neural explanation for psychological phenomena."[115] It was up to Ramachandran to complete the task. To give his analyses legitimacy, Ramachandran simply had to "anchor the airy abstractions of Freudian psychology in the physical flesh of the brain."[116]

In his efforts to use psychoanalysis to inform modern neuroscience conceptually, Ramachandran did not stray too far from a group of neuroscientists, such as the Nobel laureate Eric Kandel and the neuropsychologist and historian of psychoanalysis Mark Solms, who had created the field of "neuropsycho-analysis" (with its own journal of the same name in 1999).[117] Ramachandran's approach is interesting for us because he identified the cerebral body map as the central device for translating psychoanalysis into neuroscience. What Freud described as a psychological process was for Ramachandran a transformation at the level of the body image. Take for example Ramachandran's explanation of foot fetishes. He agreed with Freud that the foot could become symbolic of the penis, but Ramachandran linked the two through the proximity of the foot and genital area on the somatosensory map. Body parts whose cortical representation was further away from that of the genitals, as Ramachandran pointed out, were far less likely to be fetishized. To Ramachandran, the "map-dominated" character of the human being offered a far more convincing explanation than Freud's psychodynamic explanation.[118]

Similarly, Ramachandran used the body image for thinking through the Freudian conception of repression and thus of the unconscious. Taking the example of his anosognosia patients (patients who were unaware of or denied their illness), Ramachandran asserted a "striking similarity between the strategies these patients use and what Sigmund and Anna Freud called psychological defense mechanisms."[119] For Ramachandran, anosognosia and the repression it caused could be explained by the priority of the intact dynamic body image over the paralyzed body. The absolute precedence of the body map in anosognosia patients required them to deny or "repress" evidence to the contrary.[120] Ramachandran even extended the model to account for gender dysphoria. He proposes that the brains of transgender people are hardwired like brains of the other sex, which is why phantom breasts or phantom penises occurred less frequently after gender reassignment surgery than in breast and penis amputations for other reasons.[121]

That the body image was Ramachandran's central tool while grappling with Freud is telling. As we saw, Ramachandran increasingly understood the body image as a type of mirror image, produced and tinkered with through the action of mirror neurons and open to manipulation through the mirror box. And though the mirror was for Ramachandran unequivocally a purely materialist construction—mirror images could be explained by the laws of optics; there was not a ghost or phantom in sight—it remained, for him, "deeply enigmatic."[122] Despite its materiality it produces a "virtual reality," a nonmaterial doubling of the real body, that looks and even acts like a phantom. And this virtual reality is not simply a faithful copy. As for Lacan, Carpenter, and Bruch, it could be, perhaps always is, slightly out of joint with reality. The disjuncture between reality and the mirror image was crucial for Ramachandran's work: therapeutically in the mirror box it provides the image of an arm that is not, in fact, there; conceptually in the role of mirror neurons, it stands in the liminal ground between mental action and real execution.

Though the mirror map was used by Ramachandran as a way to "anchor" Freud's "airy abstractions" in the "physical flesh of the brain," it could also be interpreted in a diametrically opposed fashion. In relying so heavily on mirrors Ramachandran did not bypass the problems of the ambiguous and complex relationship between the psychological and the somatic; rather he integrated that ambiguity into the center of this somatic theory. For Ramachandran the cerebral self was a hall of mirrors, where the real and the unreal, the object and its image can never be absolutely and reliably distinguished.

Conclusion

FAILING THE TEST

AT THE BEGINNING of this book, I shared my path to this project, including both my captivation by my twin girls as they played with their mirror images, and the responses of friends and colleagues to my ideas, which suggested that mirrors tap into something deeply personal. I would like to end the book by sharing another experience, this time as a scholar of the mind and brain sciences. My first book, *Localization and Its Discontents: A Genealogy of Psychoanalysis and the Neuro Disciplines*, grappled with the relationship between the study of mental phenomena and that of the brain, by tracing the personal, institutional, and intellectual connections linking figures who, in our conventional understanding at least, seem worlds apart. This work suggested an unexpected proximity between the psychoanalysis of Sigmund Freud and the neurosurgery of Otfrid Foerster, and between the psychological body schema of Paul Schilder and brain maps of Wilder Penfield. In turn it suggested a different typology that cut across the mind/brain divide, based on different practices for intervening on the nervous system, and which promoted different ideas of the self: self-opaque and fragmented on the one hand; self-transparent and unified on the other.[1]

Once the book was finished, I decided that I wanted to study something new, and ask different questions. In the mirror project, the novel aspects held my interest: I enjoyed the concrete simplicity of the mirror test, the often playful character of the experiments, and, frankly, the fact that I could discuss babies and animals doing funny things in front of their reflections. I appreciated the ways in which the research led me into hitherto unexplored disciplinary contexts—beyond developmental psychology and psychoanalysis, I was excited to read about Enlightenment science, cybernetic circuits, media theory, and anthropology. I quickly realized that the mirror was not always as harmless and playful as it appeared at first sight. I was not simply dealing with middle-class children and cute puppies, but also with deeply troubled patients whose therapy leaned into the objectification that caused their illness in the first place; I encountered the study of non-Western ethnic groups who were the subject of the prurient gaze of American scientists, who were ultimately concerned only with their own history, and who, even in their own estimation, caused far-reaching cultural and social damage as a result.

There was also, however, a more familiar presence, which became increasingly visible as I sought to draw together the various strands of the mirror tradition, and which brought me back to the themes that had occupied me in my first book. For as I started to examine the persistent reemergence of the mirror in such a range of different disciplinary contexts, as I tried to discern that hidden force that drew so many, often independently of each other, back to reflecting surfaces, it occurred to me that the power of the mirror, for good or for ill, is that it produces something that resembles the constitutive problematic of the mind and brain sciences: how are we to understand the relationship between consciousness and matter?

The mirror might have on occasion found its place in some of the more peripheral topics in the mind and brain sciences. Especially in the 1920s and 1930s, it had a relatively diminished status, serving only to check a range of secondary cognitive abilities. And for Grey Walter, it was good for a knowing joke. But these were, we saw, in many ways falls from grace, penance for having reached too high and promised too much. Through most of its history, the mirror test has been used to pose questions that reached right to the heart of human existence. For Darwin it was related to reason and the association of ideas. For Preyer it offered a way to probe our highest foundational concept. For Carpenter it spoke to something essential about sociability. For most before Gallup, it helped define what made humans special and different from the rest of the animal kingdom. However much he may have shifted the meaning of the concept, Lacan had it right when he asserted that there is something specular about the human ego. The mirror not only seems to provide access to this self. As we saw in the final two chapters, it also offers a means to intervene. Through the mirror we can shape the self, repair it when it is broken, and realign it with reality. As philosophers of science have suggested, things become real not when we think or perceive them, but rather when they become the object of our manipulation.[2] In this way, both mirror and mirror box therapy have legitimized versions of the mirror self.

This is perhaps not the self that we read about the in the large-scale histories that have a recurrent vogue, and that tend to double up as histories of philosophy.[3] Rather it is a self that has even more authority because it is based on the sciences that in modern times have had a virtual monopoly on public perceptions of the question: psychology and, increasingly over the second half of the twentieth century, the neurosciences, broadly defined. The mirror is attractive precisely because it promises to bridge the two, much as it did for Ramachandran, for it recreates a dualism similar to many solutions of the mind-brain problem. On one side of the mirror we have a body in the flesh. On the other side, we have an immaterial world. Though they do not touch and in fact they are kept rigorously separate, there is a strict parallelism between the two, with every action in the real world tracked by another in the mirror world. Thus the mirror produces the strange impermeable and yet transparent boundary between the real and the imaginary, the neurological and the psychological, the material and the immaterial. And for scientists on both sides of the soma/psyche divide, the mirror has served as a means to understand something that otherwise lay beyond their grasp, something in that strange space between the two. a neurological self, or perhaps a virtualized brain. That is, I think, why it has generated so much interest in so many different fields over the past 150 years.

The appeal of mirrors is perhaps especially strong for the neuroscientific tradition that is now dominant. Neuroscience has been instrumental to our current tendency to identify our essential selves with the brain, to see ourselves as "cerebral subjects" as Fernando Vidal and Francisco Ortega have it in their recent book *Being Brains*.[4] This explains the enormous power of the concept of "brain death" to determine the true end of life.[5] It explains our repugnance at forms of treatment that earlier were relatively common and accepted, such as lobotomy.[6] And it explains the attraction of certain forms of transhumanism that have it that we could one day upload our neural circuits to a computer and live disembodied lives.[7] From this materialist perspective the mirror seems to have achieved an improbable balancing act. Despite its seeming concession to dualism, the mirror setup does not require one to accept a dualistic ontology. There are here no quasi-religious preconceptions required, no break from the somatic schema of modernity.

Does the mirror solve "the" mind-body problem? This is not the place to intervene in a philosophical debate that produces scores of books each year, and I certainly don't pretend to give a full answer.[8] There are reasons, however, to be skeptical. Whatever the attractive similarities between the mirror setup and the consciousness-matter relation, even an only cursory inspection reveals profound differences. Most modern conceptions of the mind-body problem restrict it more or less exclusively to the mind-*brain* problem. But of course the brain is entirely absent in the mirror encounter, because it is encased and thus hidden in our skulls. The parallelism of the mirror situation thus does not deal with thoughts and cerebral processes. So too the mirror image has few of the characteristics we might want to attribute to our minds. If anything it reflects, not ourselves but our bodies. As Wallon pointed out, self-consciousness and mirror recognition are not the same thing. In any case, of the pair, the mirror image is not the ground of will and reason; it is the passive partner, controlled down to its smallest particular by the actions of the physical body it reflects. Last but not least, the image is, in a very concrete sense, not us. Descartes worried how we could find a way from thinking substance to our bodies, and he famously offered the pineal gland as a bridge. We on the other hand are stranded in the material world, forever separated from our reflections, by the glass of the mirror.

Where does that leave the mirror-test tradition? It helps explain why the more ambitious projects to which it has been turned have tended to fall by the wayside. The mirror test no longer stands as the ultimate demarcator policing the border between humans and nonhuman animals; rather it has had a more successful career after Gallup in downplaying the differences between the two. Even before Gallup, the vaunted language of the "Ichbegriff" turned out to be the most fragile aspect of Preyer's work, at first discarded by many of his American successors, and then, when it returned, hollowed out by figures like Lacan. Carpenter's attempt to cast the mirror in the place of the foundational step in a grand historical theory of media did not survive even his own scrutiny. The more recent mirror projects have not been more successful. After much hype, the mirror neuron research program has lost much of its earlier sheen.[9] Ramachandran's work might make a good TED talk, but it hasn't found significant traction in the more established areas of academic neuroscience. Zazzo's judgment still holds: mirror research is a "domain where the ideas are so much richer than the facts."[10]

We should look instead for the legacy of the mirror self-recognition test in those of its products that at first appeared as side effects, but which have lasted longer and been more influential. For Preyer and others, the engagement with the mirror test instigated a new articulation of the relationship between psychology and neurology. In grappling with the ambiguity of mirror responses in the late nineteenth century, researchers developed new methods and struggled with the question what it meant to be scientific in psychology, at a time when the field was young and those questions were pressing. Later the mirror response presented a situation that refused any simplistic either/or between behavior and cognition and thus helped psychologists think through ways to bring them together. Moreover, it has encouraged an experimentation with different forms of media, with Gesell, Zazzo, and others, that has enriched the study of psychology. Finally, whatever problems it may still provide, mirror therapy treatment, whether for anorexics or for amputees, has transformed the lives of many people.

For these reasons, while the mirror does promise to solve or at least bypass the mind-body problem, but ultimately fails, we can still see that failure as enormously productive, one that has led scientists to think through some of the biggest questions about human identity, and that has shaped the history of the mind and brain sciences over the past 150 years. Though when we look into the mirror, we know that what we are seeing is an illusion, a trick of the light, it is still one that can fascinate and instruct.

ACKNOWLEDGMENTS

THIS BOOK BENEFITED from the insights and assistance of numerous people. I would like to thank those who provided feedback on parts of or on the whole manuscript: Edward Baring, Graham Burnett, Angela Creager, Yael Geller, Michael Gordin, Martha King, Ruth Leys, Erika Milam, Susan Sugarman, and Keith Wailoo. Their astute comments improved the book considerably. My special thanks go to Edward Baring, Michael Gordin, Martha King, and Erika Milam, who have helped me think through too many aspects of this book to list. I am deeply grateful for their insights.

I can hardly do justice to the many people who offered thoughts, provided references, suggested leads, and pointed out sources throughout the long gestation of this book, but they include Alexandra Bacopoulos-Viau, Joshua Bauchner, Cornelius Borck, Peter Brown, the late John Burnham, Jimena Canales, Stephen Casper, Ellen Chances, Hannah-Louise Clark, Sally Cochrane, Deborah Coen, Thomas Conlan, Henry Cowles, the late John Forrester, Sheldon Garon, Delia Gavrus, Cathy Gere, Stefanos Geroulanos, Jeremy Greene, Anne Harrington, Volker Hess, Daniel Hoffman-Schwartz, Stephen Jacyna, Uffa Jensen, Bill Jordan, Daniel Jütte, Markus Krajewski, Emmanuel Kreike, Howard Kushner, Susan Lanzoni, Sophie LeDebur, George Makari, Bunny McBride, Mikey McGovern, Jan Müggenburg, Barbara Nagel, Phil Nord, Scott Phelps, Harald Prins, Alicia Puglionesi, Chitra Ramalingam, Jennifer Rampling, Tobias Rees, Diane Reiss, Felix Rietmann, David Robertson, Youjung Shin, Ava Shirazi, Mark Siegeltuch, Dana Simmons, Richard Spiegel, Max Stadler, Emily Thompson, John Tresch, and Wendy Warren.

It has been a great pleasure to work with my editor at Princeton University Press, Eric Crahan, whose keen editorial eye and feeling for big questions helped me improve the manuscript considerably. I am also grateful to two anonymous reviewers and their excellent comments. At PUP, I am further indebted to Barbara Shi, who shepherded me through the process of preparing this book for publication, to Kathleen Kageff for her excellent copyediting, and to Karen Carter for managing this book's production.

I have presented my work at a number of venues and would like to thank the conveners and audiences for their helpful comments. At Princeton University, I would like to thank Joanne Gowa at the Woodrow Wilson Scholars; Phil Nord and the Work-in-Progress series in the History Department; Esther (Starry) Schor and the Behrman fellows; Candela Potente and the Theory Reading Group; Janet Currie and the Center for Health and Wellbeing; and Aaron Bornstein and the Princeton Neuroscience Institute. I am also grateful for the feedback I received from Cathy Gere and participants at the Science Studies Colloquium at the University of California–San Diego; Zachary Levine and the Neuroscience and History Lecture Series at Columbia University; Anne Hoffman and the Richardson History of Psychiatry Research Seminar at Weill Cornell Medical Center; Tabea Cornel and the "Sorting Brains Out" workshop at University of Pennsylvania; David Bates and Nima Bassiri and the "Plasticity and Pathology" workshop at UC-Berkeley; and the participants of the Soul Catchers workshops, which I organized with Volker Hess. I would also like to thank the John C. Burnham Early Career award prize committee (Ben Harris, Jill Morawski, and Laura Stark) for their helpful feedback on parts of the project.

I was lucky to have editorial and research assistance from several graduate students: Joshua Bauchner, Gabriel Lawson, David Robertson, Richard Spiegel, and Jacki Hedlund Tylor. A number of archivists and librarians provided invaluable assistance in accessing materials. I would like to thank Toni Armstrong at the Robert Hutchings Goddard Library in Worcester, Massachusetts; Caitlin J. Crennell at the Science Museum Library and Archives, Swindon, UK; Alethea Drexler at the John P. McGovern Historical Collections and Research Center at the Houston Academy of Medicine; Mark O'English at the Washington State University Archives; Christopher Hilton and Alice Mountfort at the Wellcome Library in London; Timothy H. Horning at the University Archives and Records Center, University of Pennsylvania; Paul Johnson at the National Archives at Kew; Mary Hague-Yearl and Lily Szygiel at the Osler Library at McGill University; and Sean Mooney at the Rock Foundation, New York City. Alain St. Pierre at Princeton helped more than once locate electronic copies of books, especially in the early days of COVID-19.

Two chapters of this book are based on previously published work: chapter 8 builds on an essay I published in *Pathology and Plasticity: On the Formation of the Neural Subject*, edited by David Bates and Nima Bassiri ("Imperfect

Reflections: Norms, Pathology, and Difference in Mirror Neuron Research");
and chapter 4 reworks an article published by the *Journal of the History of the
Behavioral Sciences* in 2017 ("Monkeys, Mirrors, and Me: Gordon Gallup and
the Study of Self-Recognition"). I thank the editor of the *JHBS*, Alexandra
Rutherford, for permission to reprint the work here.

I was lucky to receive funding and support from various institutions. I thank
the American Council for Learned Societies (ACLS) for awarding me a fellowship to support my sabbatical in 2018–19, and the School of Historical
Studies at the Institute for Advanced Study in Princeton for providing me with
a home during a crucial time for writing the book. At the IAS, thanks especially to Myles Jackson, Heinrich von Staden, and Brian Steininger. I would
also like to thank the former chair of my department Keith Wailoo for granting
me a full year of research leave.

And last, thanks to my family: to my parents, Helga and Reinhard; my
mother-in-law, Anstice; and the siblings, cousins, and friends (especially good
old Nina) for putting up with our extended summer visits—not seeing them
during the first COVID-19 summer reminded me how much we need the time
with them to sustain ourselves; to the twins, Sofie and Lulu, who inspired the
book thematically and conceptually; to their little sister, Kara, who entered
the world when this project was well underway, and was subjected to the occasional mirror test. The girls, as they should, took time away from my work on
this book. Writing the book also took time away from them, and I'm grateful
for their understanding and the joy they give me every day. I am most grateful,
however, to my husband, Edward Baring, for his love and support. I have
learned enormously from him, in matters of the heart as well as of the mind.

NOTES

Introduction

1. On ancient mirrors, see Ava Shirazi, "The Mirror and the Senses: Reflection and Perception in Classical Greek Thought" (PhD diss., Stanford University, 2017). The work quoted in the epigraph for the book can be found in *Borges: A Reader*, ed. Emir Rodriguez Monegal and Alastair Reid (New York: Dutton, 1981), 277–78, quote on 278.

2. The following account relies on Sabine Melchior-Bonnet, *The Mirror: A History*, trans. Katharine H. Jewett (New York: Routledge, 2001). There is a range of fascinating accounts of the history of mirrors and reflections, some popular, some more scholarly. Mark Pendergrast, *Mirror, Mirror: A History of the Human Love Affair with Reflection* (New York: Basic Books, 2003); Julian Paul Keenan, with Gordon G. Gallup Jr. and Dean Falk, *The Face in the Mirror: The Search for the Origins of Consciousness* (New York: HarperCollins, 2003); Miranda Anderson, ed., *The Book of the Mirror: An Interdisciplinary Collection Exploring the Cultural Story of the Mirror* (Newcastle, UK: Cambridge Scholars, 2008); Benjamin Goldberg, *The Mirror and Man* (Charlottesville: University Press of Virginia, 1985); Hillel Schwartz, *Culture of the Copy: Striking Likenesses, Unreasonable Facsimiles* (New York: Zone Books, 2014); Jonathan Miller, *On Reflection* (London: National Gallery Publications, 1998); Herbert Grabes, *The Mutable Glass: Mirror-Imagery in Titles and Texts of the Middle Ages and English Renaissance*, trans. Gordon Collier (Cambridge: Cambridge University Press, 1982); Rebecca Shrum, *In the Looking Glass: Mirrors and Identity in Early America* (Baltimore: Johns Hopkins University Press, 2017).

3. A Venetian mirror cost more than a painting by Raphael: eight thousand pounds as compared to three thousand. Melchior-Bonnet, *Mirror*, 30.

4. Melchior-Bonnet, *The Mirror*, 36. At the same time, they enjoyed a wealth of privileges including marrying the daughters of the nobility.

5. Melchior-Bonnet, *Mirror*, 58.

6. Melchior-Bonnet, *Mirror*, 46.

7. Justus Liebig, "Ueber Versilberung und Vergoldung von Glas," *Annalen der Chemie und Pharmacie* 98, no. 1 (1856): 132–39.

8. William H. Brock, *Justus von Liebig: The Chemical Gatekeeper* (Cambridge: Cambridge University Press, 2002), 136–37.

9. Michael Hagner, "Der Hirnspiegel und das Unheimliche," in *Röntgenportrait*, ed. Torsten Seidel (Berlin: Bühler und Heckel, 2005), 90–101, on 93.

10. Brock, *Justus von Liebig*, 139.

11. Melchior-Bonnet, *Mirror*, 97.

12. Melchior-Bonnet gives the example of France, but this was also the case elsewhere: *Mirror*, 97.

13. I owe this formulation to Keith Wailoo.

14. See especially the textbook accounts: Thomas Hardy Leahey, *A History of Modern Psychology* (Upper Saddle River, NJ: Prentice Hall, 2001); Duane P. Schultz, *A History of Modern Psychology* (New York: Academic, 1975); Morton Hunt, *The Story of Psychology* (New York: Doubleday, 1993); B. R. Hergenhahn, *An Introduction to the History of Psychology* (Belmont, CA: Wadsworth Thomson Learning, 2005); James C. Goodwin, *A History of Modern Psychology* (Hoboken, NJ: John Wiley and Sons, 2008). There are also several unusually comprehensive and nuanced overview histories of psychology, such as Wade E. Pickren and Alexandra Rutherford, *A History of Modern Psychology in Context* (New York: John Wiley and Sons, 2010); and Roger Smith, *Between Mind and Nature: A History of Psychology* (London: Reaktion Books, 2013). Smith calls psychology a "field without clear boundaries," which is evident in his approach to telling its history. Smith, *Between Mind and Nature*, 7. See also his earlier encyclopedic *The Norton History of the Human Sciences* (New York: W. W. Norton, 1997).

15. John B. Watson, "Psychology as the Behaviorist Views It," *Psychological Review* 20, no. 2 (1913): 158–77, on 158.

16. On Wundt, see especially Robert W. Rieber and David K. Robinson, *Wilhelm Wundt in History: The Making of a Scientific Psychology*, ed. Robert W. Rieber and David K. Robinson (New York: Kluwer Academic/Plenum, 2001); on the "new psychology," see esp. Lorraine Daston, "The Theory of Will versus the Science of Mind," in *The Problematic Science: Psychology in Nineteenth-Century Thought*, ed. Mitchell Ash and William Woodward (New York: Praeger, 1982), 88–115; Kurt Danziger, "The Positivist Repudiation of Wundt," *Journal of the History of the Behavioral Science* 15 (1979): 205–30; Michael Sokal, "The Origins of the New Psychology in the United States," *Physis* 43 (2006): 273–300. On reasons for the Wundt myth and a historicization of its main originator, Edwin G. Boring, see John M. O'Donnell, "The Crisis of Experimentalism in the 1920s: E. G. Boring and His Uses of History," *American Psychologist* 34, no. 4 (1979): 289–95; Alexandra Rutherford, "Maintaining Masculinity in Mid-Twentieth-Century American Psychology: Edwin Boring, Scientific Eminence, and the 'Woman Problem,'" *Osiris* 30, no. 1 (2015): 250–71. On the origins of behaviorism, see esp. John C. Burnham, "On the Origins of Behaviorism," *Journal of the History of the Behavioral Science* 4, no. 2 (1968): 143–51; John M. O'Donnell, *The Origins of Behaviorism: American Psychology, 1870–1920* (New York: New York University Press, 1985). On the origins of cognitivism, see esp. Thomas Sturm and Horst Gundlach, "Zur Geschichte und Geschichtsschreibung der 'kognitiven Revolution'—eine Reflexion," in *Handbuch Kognitionswissenschaft*, ed. Achim Stephan and Sven Walter (J. B. Metzler, 2013), 7–21. In part because disciplinary histories began to feel old and stale, recent scholarly interest has moved in new directions. See, for example, Rebecca Lemov, *World as Laboratory: Experiments with Mice, Mazes, and Men* (New York: Hill and Wang, 2005); Rebecca Lemov, *Database of Dreams: The Lost Quest to Catalog Humanity* (New Haven, CT: Yale University Press, 2015); Alexandra Rutherford, *Beyond the Box: B. F. Skinner's Technology of Behavior from Laboratory to Life, 1950s–1970s* (Toronto: University of Toronto Press, 2009); Jamie Cohen-Cole, *The Open Mind: Cold War Politics and the Sciences of Human Nature* (Chicago: University of Chicago Press, 2014); Erika L. Milam, *Creatures of Cain: The Hunt for Human Nature in Cold War America* (Princeton, NJ: Princeton University Press, 2019); Henry Cowles, *The Scientific Method: An*

Evolution of Thinking from Darwin to Dewey (Cambridge, MA: Harvard University Press, 2020); Cathy Gere, *Pain, Pleasure and the Greater Good: From the Panopticon to the Skinner Box and Beyond* (Chicago: University of Chicago Press, 2017); Howard I. Kushner, *On the Other Hand: Left Hand, Right Brain, Mental Disorder, and History* (Baltimore: Johns Hopkins University Press, 2017); Marga Vicedo, *The Nature and Nurture of Love: From Imprinting to Attachment in Cold War America* (Chicago: University of Chicago Press, 2013); Stefanos Geroulanos and Todd Meyers, *The Human Body in the Age of Catastrophe: Brittleness, Integration, Science and the Great War* (Chicago: University of Chicago Press, 2018); Deborah Weinstein, *The Pathological Family: Postwar America and the Rise of Family Therapy* (Ithaca, NY: Cornell University Press, 2013); Jan Goldstein, *The Post-revolutionary Self: Politics and Psyche in France, 1750–1850* (Cambridge, MA: Harvard University Press, 2008).

17. See Gregory Radick, *The Simian Tongue: The Long Debate about Animal Language* (Chicago: University of Chicago Press, 2007). He cites factors such as the shift of research from language to the fossil record after the discovery of Java Man in 1890, the shifting toward quantitative ideals in animal psychology, and the return of playback experiments through the doors of ethology.

18. Though less widespread, language did lend itself to a certain form of testing: Richard Garner's playback experiments, where he recorded the sound produced by great apes, and then played them back, seeing both whether he could learn to understand them, and whether the apes would. For instance, one recording, which he interpreted to be an account of the weather, when played back to the ape, led it to move to the window and look outside. See Radick, *Simian Tongue*, esp. chap. 3.

19. Adult humans, that is.

20. See Hans-Jörg Rheinberger, *Toward a History of Epistemic Things: Synthesizing Proteins in the Test Tube* (Stanford, CA: Stanford University Press, 1997). Elsewhere, Rheinberger distinguishes between "test" and "experiment" by characterizing the former as "oriented toward containing and thus closure," and the latter as "oriented toward roaming and thus aperture." He warns against the desire to make too sharp a distinction between the two however, an attitude that I adopt in this book. Hans-Jörg Rheinberger, "On Testing: An Afterword," in *Testing Hearing: The Making of Modern Aurality*, ed. Viktoria Tkaczyk, Mara Mills, and Alexandra Hui (Oxford: Oxford University Press, 2020), 351–57, on 352.

21. Robert Kohler links this wide-ranging quality to the materiality of scientific objects: "In pursuit of the history of tools and practices, historians can forage more widely and enjoy a more varied intellectual diet," specifically in contrast with intellectual and institutional history. Robert Kohler, *Lords of the Fly: Drosophila Genetics and the Experimental Life* (Chicago: University of Chicago Press, 1994), xiv. For other works that examine the continuity of material practices across different contexts ("resident science" and playback experiments), see Kohler's *Inside Science: Stories from the Field in Human and Animal Science* (Chicago: University of Chicago Press, 2019); Radick, *Simian Tongue*.

22. Peter Galison, *Image and Logic: A Material Culture of Microphysics* (Chicago: University of Chicago Press, 1997), xvii.

23. Galison, *Image and Logic*, xviii.

24. Rheinberger, *Toward a History of Epistemic Things*, 33. Note that the relationship between experimental systems and epistemic things was mutually constitutive.

25. Rheinberger, *Toward a History of Epistemic Things*, 9.

26. Rheinberger, *Toward a History of Epistemic Things*, 28.

27. Other important accounts include Steven Shapin and Simon Schaffer, *Leviathan and the Air-Pump: Hobbes, Boyle, and the Experimental Life* (Princeton, NJ: Princeton University Press, 1985); Angela Creager, *The Life of a Virus: Tobacco Mosaic Virus as an Experimental Model, 1930–1965* (Chicago: University of Chicago Press, 2002); Angela Creager, *Life Atomic: A History of Radioisotopes in Science and Medicine* (Chicago: University of Chicago Press, 2013); Kohler, *Lords of the Fly*; Karin Knorr-Cetina, "Laboratory Studies: The Cultural Approach to the Study of Science," in *Handbook of Science and Technology Studies*, ed. Sheila Jasanoff (Thousand Oaks, CA: Sage, 1994), 140–66; Michael Lynch, *Art and Artifact in Laboratory Science: A Study of Shop Work and Shop Talk in a Research Laboratory* (London: Routledge and Kegan Paul, 1985).

28. See, for example, Joseph Dumit, *Picturing Personhood: Brain Scans and Biomedical Identity* (Princeton, NJ: Princeton University Press, 2004); Anne Beaulieu, *The Space Inside the Skull: Digital Representations, Brain Mapping and Cognitive Neuroscience in the Decade of the Brain* (Amsterdam: University of Amsterdam, 2000); Kelly A. Joyce, *Magnetic Appeal: MRI and the Myth of Transparency* (Ithaca, NY: Cornell University Press, 2008); Cornelius Borck, *Hirnströme: Eine Kulturgeschichte der Elektroenzephalographie* (Göttingen: Wallstein, 2005); Weinstein, *Pathological Family*; Felix Rietmann, "Seeing the Infant: Audiovisual Technologies and the Mind Sciences of the Child" (PhD diss., Princeton University, 2018); Hagner, *Der Geist bei der Arbeit: Historische Untersuchungen zur Hirnforschung* (Göttingen: Wallstein, 2006); Hagner, "Der Hirnspiegel und das Unheimliche"; Frank Stahnisch, "The Language of Visual Representation in the Neurosciences: Relating Past and Future," *Translational Neuroscience* 5, no. 1 (2014): 78–90. For a recent overview on media studies and the history of knowledge, see Jeremy Greene, "Knowledge in Medias Res: Toward a Media History of Science, Medicine, and Technology," *History and Theory* 59, no. 4 (2020): 48–66.

29. Cornelius Borck, "Schreibende Gehirne," in *Psychographien*, ed. Cornelius Borck and Achim Schäfer (Zurich: Diaphanes, 2005), 89–110, on 109.

30. Borck and Schäfer, *Psychographien*.

31. Because humans are bilaterally symmetrical, this applies even to our own body, with the important proviso that we never see our body whole except in the mirror; see my discussion below. For a discussion of antique debates about why mirror images are inverted only left-right, not up-down, see Daryn Lehoux, "Observers, Objects, and the Embedded Eye," *Isis* 98 (2007): 447–67.

32. For recent overviews, see Sandra Pravica, "'Materialität' in der Naturwissenschaftsforschung: Eine bibliographische Übersicht" (working paper, Max-Planck-Institut für Wissenschaftsgeschichte, Berlin, 2007); Simon Werrett, "Matter and Facts: Material Culture and the History of Science," in *Material Evidence: Learning from Archaeological Practice*, ed. Robert Chapman and Alison Wylie (New York: Routledge, 2014), 339–52.

33. Ursula Klein, *Experiments, Models, Paper Tools: Cultures of Organic Chemistry in the Nineteenth Century* (Stanford, CA: Stanford University Press, 2003). See also Bernhard Dotzler, *Papiermaschinen: Versuch über Communication and Control in Literatur und Technik* (Berlin: Akademie-Verlag, 1996). This also applies to Latour and Woolgar's "inscriptions": Bruno Latour and Steve Woolgar, *Laboratory Life: The Social Construction of Scientific Facts* (Princeton, NJ: Princeton University Press, 1986).

34. See, e.g., Anke te Heesen, "The Note-Book: A Paper Technology," in *Making Things Public: Atmospheres of Democracy*, ed. Bruno Latour and Peter Weibel (Cambridge, MA: MIT Press, 2003), 582–89. Volker Hess and Andrew Mendelsohn, "Case and Series: Medical Knowledge and Paper Technology, 1600–1900," *History of Science* 48, no. 3 (2010): 582–89. For an overview of the historiography on paper technologies, see Carla Bittel, Elaine Leong, and Christine von Oertzen, "Introduction: Paper, Gender, and the History of Knowledge," in *Working with Paper: Gendered Practices in the History of Knowledge*, ed. Carla Bittel, Elaine Leong, Christine von Oertzen (Pittsburgh, PA: University of Pittsburgh Press, 2019). Emil Kraepelin's diagnostic patient cards are a good example of a paper tool in the mind sciences. These cards (containing in brief form the causes, etiology, phenomena, development, and prognosis of the disease) were used for the construction of his nosological system. But they were also saturated with economic considerations: in order to quickly move incurable patients to the peripheral hospitals (*Landesanstalten*), the diagnostic card also included a prognosis. See Eric Engstrom, *Clinical Psychiatry in Imperial Germany: A History of Psychiatric Practice* (Ithaca, NY: Cornell University Press, 2003). See also Theodore Porter's fascinating account of data tables and hereditary data collection in the nineteenth-century asylum. Theodore M. Porter, *Genetics in the Madhouse: The Unknown History of Human Heredity* (Princeton, NJ: Princeton University Press, 2018).

35. Soraya de Chadarevian and Nick Hopwood, eds., *Models: The Third Dimension of Science* (Stanford, CA: Stanford University Press, 2004); Lorraine Daston, ed., *Things That Talk: Object Lessons from Art and Science* (New York: Zone Books, 2008).

Chapter 1: My Child in the Mirror

1. Charles Darwin, Diary of an infant, CUL DAR 210.11:37, transcript at Darwin Correspondence Project, Cambridge University Library, https://www.darwinproject.ac.uk/people/about -darwin/family-life/darwin-s-observations-his-children.

2. Of course Darwin found the observations interesting precisely because they pointed to a universal pattern, beyond the particularities of his own child.

3. M notebook, 1838, CUL DAR 125, transcript at Darwin Online, http://darwin-online.org .uk/content/frameset?keywords=history%20babies%20of%20natural&pageseq=149&itemID =CUL-DAR125.-&viewtype=text.

4. Darwin thought his diary keeping contributed to a "natural history of babies." M notebook, 157, cited in Howard E. Gruber and Robert T. Keegan, "Charles Darwin's Unpublished 'Diary of an Infant': An Early Phase in His Psychological Work," in *Contributions to a History of Developmental Psychology: International William T. Preyer Symposium* (New York: Mouton, 1985), 127–45. See also Marjorie Lorch and Paula Hellal, "Darwin's 'Natural Science of Babies,'" *Journal of the History of the Neurosciences* 19, no. 2 (2010): 140–57. On Darwin's notebooks, see Charles Darwin, *Charles Darwin's Notebooks, 1836–1844: Geology, Transmutation of Species, Metaphysical Enquiries* (Ithaca, NY: Cornell University Press, 1987).

5. Charles Darwin, *The Expression of the Emotions in Man and Animals* (London: J. Murray, 1872); and Charles Darwin, "A Biographical Sketch of an Infant," *Mind* 2 (1877): 285–94.

6. Darwin, "Biographical Sketch of an Infant," 290.

7. Darwin, "Biographical Sketch of an Infant," 290.

8. As Sally Shuttleworth observes, Darwin himself did not understand the fuss about his *Mind* article and stated that it was not worthy of the attention it received, Sally Shuttleworth, *The Mind of the Child: Child Development in Literature, Science, and Medicine, 1840–1900* (Oxford: Oxford University Press, 2010), 228.

9. Johann Peter Süssmilch, *Versuch eines Beweises dass die erste Sprache ihren Ursprung nicht vom Menschen, sondern allein vom Schöpfer erhalten habe, etc.* (Berlin, 1766). The title translates to "Attempt to prove that the first language does not originate from man, but from God." To my knowledge, the work has not been translated into English.

10. "Haben die Menschen, ihrer Naturfähigkeit überlassen, sich selbst Sprache erfinden können? Und auf welchem Wege wären sie am füglichsten dazu gelangt?" Herder included both questions in his essays, one at the beginning of part 1, the other at the beginning of part 2. Johann Gottfried von Herder, "Treatise on the Origin of Language (1772)," in *Philosophical Writings*, ed. and trans. Michael N. Forster (Cambridge: Cambridge University Press, 2002), 65–164. On Süssmilch and the competition, see Allan Megill, "The Enlightenment Debate on the Origin of Language and Its Historical Background" (PhD diss., Columbia University, 1975), esp. chaps. 11, 12.

11. Herder, "Treatise on the Origin of Language," 96.

12. Herder, "Treatise on the Origin of Language," 88–89.

13. Herder, "Treatise on the Origin of Language," 99, 100.

14. Johann Gottfried v. Herder, *Outlines of a Philosophy of the History of Man*, 1785, trans. T. Churchill (1800; New York: Bergman, 1977), 233–34.

15. The text was later published: Dietrich Tiedemann, *Versuch einer Erklärung des Ursprunges der Sprache* (Riga, 1772). English: "Attempt to explain the origins of language." To my knowledge, no English translation exists. See below on the history of the text.

16. Tiedemann, *Versuch einer Erklärung des Ursprunges der Sprache*, 181.

17. In his *Beobachtungen* of 1787, while equally embracing the language barrier, Tiedemann located the difference between animals and humans in the human ability of discernment (animals lacked the higher powers of the soul [*höhere Seelenkräfte*] of judgment [*Urteil*] and comparison [*Vergleichen*], which first became visible in their use of articulate language). Dietrich Tiedemann, *Beobachtungen über die Entwickelung der Seelenfähigkeiten bei Kindern* (Altenburg: Oskar Bonde, 1897). See also Suzie Bartsch, *Dietrich Tiedemann: Funktionalistische Sprachtheorie und Sprachursprungstheorie* (Munich: GRIN, 2015).

18. On the rise and decline of the baby-diary tradition, see Doris B. Wallace, Margery B. Franklin, and Robert T. Keegan, "The Observing Eye: A Century of Baby Diaries," *Human Development* 37, no. 1 (1994): 1–29.

19. For Tetens, psychology was predominantly a means of understanding the possibilities of adult human reasoning, and his assumptions were famously developed by Immanuel Kant. On Tetens, see Roger Smith, *The Norton History of the Human Sciences* (New York: W. W. Norton, 1997), 204–6. See also Thomas Sturm, *Kant und die Wissenschaften vom Menschen* (Paderborn: Mentis, 2009).

20. In his *Allgemeine Revision des gesamten Schul- und Erziehungswesen* of 1785, he called for the composition of diaries, or "educational histories," by trained observers (that is, fathers) to capture the physical and moral development of a child from birth, an attempt to place pedagogy on a systematic foundation and to inform the establishment of a public school system. See Siegfried Jaeger, "The Origin of the Diary Method in Developmental Psychology," in

Contributions to a History of Developmental Psychology: International William T. Preyer Symposium (New York: Mouton, 1985), 63–74, on 67–68. See also Pia Schmid, "Vätertagebücher des ausgehenden 18. Jahrhunderts: Zu Anfängen der empirischen Forschung von Säuglingen und Kleinkindern," in *Kinder, Kindheit, Lebensgeschichte: Ein Handbuch*, ed. Imbke Behnken and Jürgen Zinnecker (Seelze-Velber: Kallmeyer, 2001), 325–39; Simone Austermann, *Die "Allgemeine Revision": Pädagogische Theorieentwicklung im 18. Jahrhundert: Beiträge zur Theorie und Geschichte der Erziehungswissenschaft*, vol. 32 (Bad Heilbrunn: Julius Klinkhardt, 2010). Elizabeth Cleghorn Gaskell, *Private Voices: The Diaries of Elizabeth Gaskell and Sophia Holland* (Keele, Staffordshire: Keele University Press, 1996).

21. The British novelist Elizabeth Gaskell, famous for her portrayal of social life in industrializing Britain (as well as her ghost stories), is a case in point. She started her diary on March 10, 1835, when her first child, Marianne, was six months old, and she intended to keep it private (it was published only in 1923). Her diary was clearly an attempt to work through a process that was for her associated with uncertainty and guilt.

22. In fact, Tiedemann was an outspoken empiricist and opponent of Kantian idealism. For him, language, and the development of cognitive abilities more broadly, was not an a priori subject.

23. Tiedemann's baby diary was first published in *Hessische Beiträge zur Gelehrsamkeit und Kunst 2* (1787), 313–33 and 486–502. It became better known in Germany only via the reception in France, where it was published in translation first in the *Journal général de l'instruction publique* in 1863 and then in Bernard Perez's *Thierri Tiedemann et la science de l'enfant: Mes deux chats; fragment de psychologie comparée* (Paris: Germer Baillière), in 1881. Christian Ufer, introduction to *Beobachtungen über die Entwickelung der Seelenfähigkeiten bei Kindern*, by Dietrich Tiedemann (Altenburg: Oskar Bonde, 1897), v. English editions followed: Dietrich Tiedemann, *Tiedemann's Record of Infant-Life: An English Version of the French Translation and Commentary by Bernard Perez*, notes by F. Louis Soldan (Syracuse, NY: C. W. Bardeen, 1890); Carl Murchison and Suzanne Langer, "Tiedemann's Observations on the Development of the Mental Faculties of Children," *Pedagogical Seminary and Journal of Genetic Psychology* 34, no. 2 (1927): 205–30. This last translation was reprinted as Dietrich Tiedemann, "Observations on the Mental Development of a Child (1787)," in *Historical Readings in Developmental Psychology*, ed. Wayne Dennis (New York: Appleton-Century-Crofts, 1972), 11–31. For Tiedemann's larger philosophical system, where he drew on the observations recorded in the baby diary, see Dietrich Tiedemann, *Handbuch der Psychologie, zum Gebrauche bei Vorlesungen und zur Selbstbelehrung bestimmt*, ed. Ludwig Wachler (Leipzig: Barth, 1804), 401–31.

24. Tiedemann, *Beobachtungen*, 20.

25. Tiedemann, *Beobachtungen*, 20–21.

26. Tiedemann, *Beobachtungen*, 23–24.

27. Tiedemann, *Beobachtungen*, 23.

28. Tracking the development of his own son, he saw the first germs of language "in the middle of the first quarter," which often frightened his mother who believed "she heard an angel's voice" when the child began to babble single syllables, "Ma, Ba, Bu." Only later did he begin to imitate sounds, first musical tones, then the sounds of sneezing, and finally words. The first spontaneously uttered word was not "Mama" or "Papa," as his parents might have hoped for, but "Neuback" (freshly baked), picked up from the bread boy (*Brezeljunge*), who used to shout

it in the street. Berthold Sigismund, *Kind und Welt: Vätern, Muttern und Kinderfreunden* (Braunschweig: F. Vieweg und Sohn, 1856), 28, 122.

29. Taine, together with the psychologist Théodule Ribot, was part of a new secular orientation among French scholars who opposed the *spiritualisme* of a unified soul promoted by both Catholicism and the dominant Cousinian psychology. They drew inspiration from German experimental methods, advocating for a secular science of the mind. See Hippolyte Taine, *De l'intelligence* (Paris: Hachette, 1870). For a useful overview of national styles in nineteenth-century psychology, see Roger Smith, *Between Mind and Nature: A History of Psychology* (London: Reaktion Books, 2013), esp. chap. 3.

30. For an excellent account of the Müller-Darwin debate, see Gregory Radick, *The Simian Tongue: The Long Debate about Animal Language* (Chicago: University of Chicago Press, 2007).

31. Max Müller, "Lectures on Mr. Darwin's Philosophy of Language: Second Lecture," *Fraser's Magazine*, no. 42 (June 1873): 659–78.

32. Müller, "Lectures on Mr. Darwin's Philosophy of Language: Second Lecture," 677, 678.

33. Quoted in Radick, *Simian Tongue*, 16.

34. Max Müller, "Lectures on Mr. Darwin's Philosophy of Language: Third Lecture," *Fraser's Magazine* 8, no. 43 (July 1873): 1–24, on 23.

35. Müller, "Lectures on Mr. Darwin's Philosophy of Language: Second Lecture," 678. The term "radical" refers to Müller's theory of the "roots" of language. Müller's argument formed part of a broader concern for the monogenesis of the human species.

36. Müller, "Lectures on Mr. Darwin's Philosophy of Language: Third Lecture," 17.

37. Müller, "Lectures on Mr. Darwin's Philosophy of Language: Third Lecture," 17, 18.

38. Müller, "Lectures on Mr. Darwin's Philosophy of Language: Third Lecture," 19.

39. Gruber and Keegan argue that Taine had "read Müller's critique of Darwin and rejected it. Taine presented a case of the natural origin and development of language *in a child* by means of his own activity." Howard E. Gruber and Robert T. Keegan, "Charles Darwin's Unpublished 'Diary of an Infant': An Early Phase in His Psychological Work," in *Contributions to a History of Developmental Psychology: International William T. Preyer Symposium* (New York: Mouton, 1985), 127–45, on 129. But it is clear that this is a misreading of Taine's account.

40. Hippolyte Taine, "Note sur l'acquisition du langage chez les enfants et dans l'espèce humaine," *Revue philosophique de la France et de l'étranger* 1 (1876): 5–23, on 15. I cite from the English version of Taine's article, "M. Taine on the Acquisition of Language by Children," *Mind* 2, no. 6 (1877): 252–59, wherever possible. I translate some passages from the French for reasons that will become clear later in the chapter; those passages refer to the French original.

41. Taine, "M. Taine on the Acquisition of Language by Children," 253.

42. Taine, "M. Taine on the Acquisition of Language by Children," 253.

43. Taine, "M. Taine on the Acquisition of Language by Children," 252.

44. Taine, "M. Taine on the Acquisition of Language by Children," 254.

45. Taine, "M. Taine on the Acquisition of Language by Children," 257.

46. Taine, "M. Taine on the Acquisition of Language by Children," 255.

47. Taine, "Note sur l'acquisition du langage," 22, 23.

48. Taine, "Note sur l'acquisition du langage," 22.

49. Taine, "M. Taine on the Acquisition of Language by Children," 259.

50. Darwin, "Biographical Sketch of an Infant," 285. Lorch and Hellal, "Darwin's 'Natural Science of Babies,'" 142. Darwin's M and N notebooks included reflections on children and the development of language and behavior.

51. Darwin, "Biographical Sketch of an Infant," 293.

52. Darwin, "Biographical Sketch of an Infant," 294; see also 290. Darwin seems most interested in the association of ideas, rather than (as for Taine and Müller) the association of words and concepts.

53. Darwin, "Biographical Sketch of an Infant," 294.

54. Charles Darwin, *The Descent of Man* (London: John Murray, 1871), 130.

55. Darwin, *Descent of Man*, 86.

56. Darwin, "Biographical Sketch of an Infant," 290.

57. Dietrich Tiedemann, "Beobachtungen über die Entwicklung der Seelenfähigkeiten bei Kindern," *Hessische Beiträge zur Gelehrsamkeit und Kunst* 2 (1787): 313–33, 486–502.

58. Perez, *Thierri Tiedemann*. Note that "Thierri" was Preyer's middle name; Tiedemann's first name was Friedrich. The German second edition was motivated precisely by the wish to offer the reader the unmediated text. Tiedemann, *Beobachtungen*, vii. It was translated into English at the same time: Dietrich Tiedemann, *Tiedemann's Record of Infant-Life* (Syracuse, NY: C. W. Bardeen, 1890).

59. Perez, *Thierri Tiedemann*, v.

60. Perez, *Thierri Tiedemann*, 22–25. Perez did not think that there was a physiological difference that allowed the child to imitate such sounds.

61. Bernard Perez, *La psychologie de l'enfant: Les trois premières années* (Paris: Germer Baillière, 1882), 276.

62. This did of course not mean that language was entirely lost as a boundary marker. In fact, it kept coming back, both at the time of Darwin and later, as we will see later in this book. See also Radick, *Simian Tongue*.

63. Immanuel David Mauchart, "Tagbuch über die allmähliche körperliche und geistige Entwickelung eines Kindes," *Allgemeines Repertorium für empirische Psychologie und verwandte Wissenschaften* 4 (1798): 269–94.

64. Mauchart, "Tagbuch," 293.

65. In the section "The Child at Nine Months." Emma Willard, "Observations upon an Infant during Its First Year: By a Mother," appendix to *Progressive Education, Commencing with the Infant*, by Albertine Necker de Saussure, trans. and ed. Emma Willard and Almira Hart Lincoln (Boston: W. D. Ticknor, 1835): 323–48, on 336–40.

66. Sigismund, *Kind und Welt*, 54.

67. Sigismund, *Kind und Welt*, 54. It is difficult to know, of course, what joy looks like in animals.

68. Tiedemann, *Beobachtungen*, 24.

69. Tiedemann, *Beobachtungen*, 28.

70. Tiedemann, *Beobachtungen*, 24, 29.

71. Tiedemann, *Beobachtungen*, 28–29.

72. See also Tiedemann's treatise on language (*Versuch einer Erklärung des Ursprung der Sprache*), where he had argued that animals did not possess language (see above). He also wrote on the definition of "word" as connecting with *Vorstellung*: animals and infants and certain

adults utter sounds without connection—so those who grant language to animals should reconsider their definition of "word." Tiedemann, *Versuch einer Erklärung des Ursprung der Sprache*, 20–21.

73. Perez, *Thierri Tiedemann*, 31.

74. See Perez, *La psychologie de l'enfant*, chap. 13.

75. Wilhelm Preyer, *Die Seele des Kindes: Beobachtungen über die geistige Entwickelung des Menschen in den ersten Lebensjahren* (Leipzig: Th. Grieben, 1895). The first edition was in 1882. The English translation is *The Mind of the Child* (New York: Arno, 1973). All translations from the German are mine.

76. The secondary literature on Preyer is relatively sparse. The best accounts are Georg Eckardt, "Einleitung," in *Die Seele des Kindes: Eingeleitet und mit Materialien zur Rezeptionsgeschichte versehen von Georg Eckardt* (Berlin: Springer, 1989): 11–52; Siegfried Jaeger, "Origins of Child Psychology: William Preyer," in *The Problematic Science: Psychology in Nineteenth-Century Thought*, ed. William R. Woodward and Mitchell G. Ash (New York: Praeger, 1982), 300–321; Frank Richter, "Der Physiologe William Thierry Preyer (1841–1897): Dem Darwinismus verpflichtet," in *Wegbereiter der modernen Medizin*, ed. Christian Fleck (Jena: Bussert und Stadeler, 2004), 169–82.

77. Sadly, the original diary seems to have been lost.

78. Eckardt, "Einleitung," 26. In his biography of Darwin, Preyer reported that he "las es [*Origin of Species*], las es wieder und wieder und war von dem Inhalt geradezu überwältigt. . . . Alles Lebende, . . . der Zusammenhang der Naturvorgänge untereinander und das Verhältnis des Menschen zu ihnen, gewannen ein anderes Ansehen." Eckardt, "Einleitung," 26. See also Wilhelm Preyer, "Briefe von Darwin," *Deutsche Rundschau* 67 (1891): 356–90.

79. Preyer, "Briefe von Darwin." Preyer described a "popular lecture on the struggle for existence" given to an audience of six hundred at Cologne and a lecture course, "Ueber die Darwinsche Theorie," held at the University of Bonn, which attracted over two hundred students; the former was later published as "Der Kampf um das Dasein" in *Aus Natur- und Menschenleben*, by Wilhelm Preyer (Berlin: Allgemeiner Verein für Deutsche Literatur, 1885), 1–38. Preyer to Darwin, March 21, 1869, in "Briefe von Darwin." Preyer also wrote a couple of biographical accounts of Darwin: Wilhelm Preyer, "Charles Darwin: Eine biographische Skizze," *Das Ausland* 43, no. 14 (April 2, 1870): 314–20; Wilhelm Preyer, *Darwin: Sein Leben und Wirken* (Berlin: E. Hofmann, 1896).

80. A turn of phrase by the contemporaneous zoologist Carl Vogt, who used it polemically. Preyer himself emphasized Jena's place in one of his letters: "Besides Jena there is no University in Germany where your theory is so openly confessed and publicly taught by so many professors. Häckel, Gegenbaur, Dohrn, Strasburger, W. Müller, myself: we are true Darwinians, in our lectures and writings." Preyer to Darwin, April 27, 1871, Darwin Correspondence Project. Preyer referred to Ernst Haeckel, Calr Gegenbaur, Anton Dohrn, Adolf Strasburger, and Wilhelm Müller, all of whom were at the University of Jena.

81. Wilhelm Preyer, review of *Grundzüge der physiologischen Psychologie*, by Wilhelm Wundt, *Jenaer Literaturzeitung* 1, no. 5, 36 (1874): 71–72, 550–51.

82. Ernst Haeckel, *Generelle Morphologie der Organismen* (Berlin: G. Reimer, 1866).

83. Darwin participated in this tradition too. His *Descent of Man* (1871) contained a chapter on embryology. The argument that the development of embryos illustrates the history of animal

species can be traced back to Jean-Baptiste Lamarck, Etienne Geoffroy Saint-Hilaire, and Robert Chambers.

84. Ernst Haeckel, *Anthropogenie oder Entwickelungsgeschichte der Menschen* (Leipzig: Engelmann, 1874), 704. Cited in Eckardt, "Einleitung," 17.

85. Wilhelm Preyer, *Specielle Physiologie des Embryo: Untersuchungen ueber die Lebenserscheinungen vor der Geburt* (Leipzig: Th. Grieben, 1885).

86. Preyer to Darwin, July 6, 1877, Darwin Correspondence Project. Preyer was born and spent his childhood in England, which explains his proficiency in the language.

87. Preyer, "Briefe von Darwin," 378. This concurred with Darwin's own judgment of his own work. Shuttleworth, *Mind of the Child*, 228.

88. Preyer, *Darwin*, 122.

89. Preyer, *Die Seele des Kindes* (1895), v–vi.

90. Wilhelm Preyer, *Die geistige Entwickelung in der ersten Kindheit, nebst Anweisungen für Eltern, dieselbe zu beobachten* (Stuttgart: Union Deutsche Verlagsgesellschaft, 1893), 141. It is not clear if this was the method Preyer actually followed. But it likely is because he recommended it to the readers of his book.

91. Preyer, *Die Seele des Kindes* (1895), vi.

92. Wilhelm Preyer, *Die Seele des Kindes* (1895), 290–91 (my emphasis). There is a very similar quote in *Die geistige Entwickelung*, 73.

93. Note that Preyer devotes a substantial portion of his book to the discussion of language development.

94. Preyer, *Die Seele des Kindes* (1895), 389.

95. Preyer, *Die Seele des Kindes* (1895), 386.

96. Preyer, *Die Seele des Kindes* (1895), 387.

97. Preyer, *Die Seele des Kindes* (1895), 387.

98. Preyer, *Die Seele des Kindes* (1895), 387.

99. Preyer, *Die Seele des Kindes* (1895), 387.

100. Preyer, *Die Seele des Kindes* (1895), 387.

101. Throughout Preyer's book, the gender of the child did not seem to matter: He usually referred to the subject as "the newborn" (*das Neugeborene*), "the infant" (*der Säugling*, which is grammatically male but neutral as to the gender of the baby), or "the child" (*das Kind*). Even when referring to the children from other people's diaries, such as Taine's daughter, or Darwin's and Sigismund's sons, Preyer mostly stuck to the neuter *das Kind*. It was only at a stage of life when the mirror development was largely complete that Preyer considered traditional gender norms to kick in again, which explains both Preyer's desire and his willingness to withdraw his son from the mirror when he showed signs of vanity.

102. Preyer, *Die Seele des Kindes* (1895), 388.

103. Preyer, *Die Seele des Kindes* (1895), 388.

104. Preyer, *Die Seele des Kindes* (1895), 388.

105. Maximilian Schmidt, "Beobachtungen am Orang-Utan," *Der zoologische Garten* 19, no. 7, 8 (1878): 193–98, 226–33, on 232.

106. Schmidt, "Beobachtungen am Orang-Utan," 232.

107. Johann von Fischer, "Aus dem Leben eines jungen Mandril (*Cynocephalus mormon*): Seine Erkrankung und sein Tod," *Der Zoologische Garten* 17, no. 4 (1876): 116–27, on 119.

108. Charles Darwin, "Sexual Selection in Relation to Monkeys," *Nature* 15 (November 2, 1876): 18–19. Darwin did not, however, discuss the implied homosexuality of the behavior. There is one much earlier, pre-Darwinian account of an orangutan's behavior in front of the mirror. The Calcutta surgeon J. Grant, in his detailed account in 1828 of the animal's appearance and behavior, noted that the ape showed no interest whatsoever in the looking glass, something that Grant noted as surprising given the animal's general curiosity, but did not further comment on. I consider this study an outlier that can be explained by the greater access to apes and monkeys in India. J. Grant, "Account of the Structure and Habits of an Orang Outang from Borneo," *Edinburgh Journal of Science* 9 (1828): 1–25.

109. Preyer, *Die Seele des Kindes* (1895), 388.

110. Preyer, *Die Seele des Kindes* (1895), 390.

111. Preyer, *Die geistige Entwickelung*.

112. Preyer, *Die geistige Entwickelung*, 187.

113. Preyer, *Die Seele des Kindes* (1895), 390.

114. On Spalding, see Philip Howard Gray, "Spalding and His Influence on Research in Developmental Behavior," *Journal of the History of the Behavioral Sciences* 3, no. 2 (1967): 168–79.

115. Preyer, *Die Seele des Kindes* (1895), 48.

116. Throughout his book, Preyer does not give an example for a concept in animals. This was probably because he just conceded it to Darwin (and thus did not feel the need to prove the point that animals had concepts).

117. Preyer, *Die Seele des Kindes* (1895), 397.

118. Preyer, *Die Seele des Kindes* (1895), 390.

119. Preyer, *Die Seele des Kindes* (1895), 384, 432.

120. Preyer, *Die Seele des Kindes* (1895), 391.

121. Preyer, *Die Seele des Kindes* (1895), 385.

122. Preyer, *Die Seele des Kindes* (1895), 385.

123. Preyer, *Die Seele des Kindes* (1895), 392.

124. See George John Romanes, *Mental Evolution in Man: Origin of Human Faculty* (London: Kegan Paul, Trench, 1888), chap. 10 ("Self-Consciousness"). An animal knows that the grass is green, but only we "know that we know that the grass is green." Romanes gives two definitions of self-consciousness in his text: "The faculty of separating in thought the ego from the non-ego" (77), and the ability "to compare its [the mind's] past with its present, and so to reach that apprehension of continuity among its own states wherein the full introspective consciousness of self consists" (206).

Chapter 2: "Not Suddenly, but by Degrees"

1. John E. Anderson, foreword to *The First Two Years: A Study of Twenty-Five Babies*, by Mary Shirley, 3 vols. (Minneapolis: University of Minnesota Press, 1931–33), 1:vi.

2. We see a similar phenomenon in other public health movements such as eugenics, birth control, and temperance. See Greta Jones, "Eugenics and Social Policy between the Wars," *Historical Journal* 25, no. 3 (1982): 717–28; Wendy Kline, *Building a Better Race: Gender, Sexuality, and Eugenics from the Turn of the Century to the Baby Boom* (Berkeley: University of California

Press, 2005); Richard A. Soloway, "The 'Perfect Contraceptive': Eugenics and Birth Control Research in Britain and America in the Interwar Years," *Journal of Contemporary History* 30 (1995): 637–64.

3. As we will see, the relationship of the "new psychologists" to Wundt is complex. In their appropriation of Wundtian principles, many of his students moved significantly beyond him.

4. Elizabeth Stow Brown, "The Baby's Mind: Studies in Infant Psychology," *Babyhood* 68 (1890): 239–42, 274–76, 305–7, 340–42, 369–72, on 241.

5. Brown, "Baby's Mind," 341.

6. Frederick Tracy, *The Psychology of Childhood* (Boston: D. C. Heath, 1893).

7. Tracy, *Psychology of Childhood*, 47.

8. James Sully, *Studies of Childhood* (New York: Longmans, Green, 1896), 5.

9. David R. Major, *First Steps in Mental Growth: A Series of Studies in the Development of Infancy* (New York: Macmillan, 1906), 2.

10. Major, *First Steps in Mental Growth*, 124.

11. Charles Darwin, Diary of an infant, CUL DAR 210.11:37, transcript at Darwin Correspondence Project, Cambridge University Library, https://www.darwinproject.ac.uk/people/about-darwin/family-life/darwin-s-observations-his-children. Ernst Scupin and Gertrud Scupin, *Bubis erste Kindheit: Ein Tagebuch über die geistige Entwicklung eines Knaben während der ersten drei Lebensjahre* (Leipzig: Th. Grieben, 1907), 38; Wilhelm Preyer, *Die Seele des Kindes: Beobachtungen über die geistige Entwickelung des Menschen in den ersten Lebensjahren* (Leipzig: Th. Grieben, 1895), 387.

12. Charles Darwin, Diary of an infant, CUL DAR 210.11:37; Bernard Perez, *Thierri Tiedemann et la science de l'enfant: Mes deux chats; fragment de psychologie comparée* (Paris: Germer Baillière, 1881), 31.

13. Scupin and Scupin, *Bubis erste Kindheit*, 53; Sully, *Studies of Childhood*, 113; Major, *First Steps in Mental Growth*, 273.

14. Berthold Sigismund, *Kind und Welt: Vätern, Muttern und Kinderfreunden* (Braunschweig: F. Vieweg und Sohn, 1856), xi.

15. In his talk "Psychogenesis" of 1880, Preyer noted: "Hierbei wäre zu wünschen, dass mehrere physiologisch gründlich unterrichtete Männer unabhängig voneinander eine grössere Anzahl von Neugeborenen und Säuglingen sorgfältig beobachteten und die erhaltenen Resultate verglichen." Wilhelm Preyer, "Psychogenesis," in *Naturwissenschaftliche Tatsachen und Probleme* (Berlin: Paetel, 1880), 199–238, on 204. Preyer also responded positively to Emily Talbot's project, discussed below.

16. Preyer, *Die Seele des Kindes* (1895), vi.

17. Preyer, *Die Seele des Kindes* (1895), vi.

18. Wilhelm Preyer, *Die geistige Entwickelung in der ersten Kindheit, nebst Anweisungen für Eltern, dieselbe zu beobachten* (Stuttgart: Union Deutsche Verlagsgesellschaft, 1893), v–vi, vi.

19. Preyer, *Die geistige Entwickelung*, 142. This implied a distinction not unlike William Whewell's between "scientists" (his coinage) and data gatherers. On Whewell's use of the term *scientist*, see the classic paper by Sydney Ross, "Scientist: The Story of a Word," *Annals of Science* 18 (1962): 65–85. As we will see, Preyer accepted as adequate the collected data of mothers and other female relatives, though not nannies or nurses, showing how considerations of class, as well as gender, informed his research practices.

20. James Sully, notes, *Mind*, n.s., 2 (1893): 420–21.

21. On observation and collective empiricism, see Lorraine Daston and Peter Galison, *Objectivity* (New York: Zone Books, 2007); see also Lorraine Daston, "The Empire of Observation, 1600–1800," in *Histories of Scientific Observation*, ed. Lorraine Daston and Elizabeth Lunbeck (Chicago: University of Chicago Press, 2011), 81–113. On women and science, see esp. Margaret Rossiter, *Women Scientists in America: Struggles and Strategies to 1940* (Baltimore: Johns Hopkins University Press, 1982); Kimberly Hamlin, *From Eve to Evolution: Darwin, Science, and Women's Rights in Gilded Age America* (Chicago: University of Chicago Press, 2014); Sally Gregory Kohlstedt, "In from the Periphery: American Women in Science, 1830–1880," *Signs* 4, no. 1 (1978): 81–96; Mary Terrall, "Émilie du Châtelet and the Gendering of Science," *History of Science* 33, no. 3 (1995): 283–310.

22. Emily Talbot, "Report of the Secretary of the Department," in *Papers on Infant Development: Published by the Education Department of the American Social Science Association, January, 1882*, ed. Emily Talbot (Boston: Tolman and White, 1882), 5–6, on 5.

23. Emily Talbot, "Register of Infant Development: Circular of April, 1881," in E. Talbot, *Papers on Infant Development*, 49.

24. E. Talbot, "Report," 6.

25. It also built on a number of precedents for crowdsourcing projects, some of which involved mailed questionnaires and surveys, e.g., William Whewell's Great Tide Experiment of 1935. On the collective activities of Victorian naturalists, the Jesuits, and explorers in the Spanish Empire, see Jim Endersby, *Imperial Nature: Joseph Hooker and the Practices of Victorian Science* (Chicago: University of Chicago Press, 2008); Daniela Bleichmar, *Visible Empire: Botanical Expeditions and Visual Culture in the Hispanic Enlightenment* (Chicago: University of Chicago Press, 2012); Florence Hsia, *Sojourners in a Strange Land: Jesuits and Their Scientific Missions in Late Imperial China* (Chicago: University of Chicago Press, 2009).

26. Emily Talbot, "Register of Infant Development: Circular of January, 1882," in E. Talbot, *Papers on Infant Development*, 50.

27. Annie B. Howes, "The Study of the Development of Children," *ACA Journal* 4 (1891): 1–10, on 9. Howes was writing against the prejudices held by people such as Sully against female observers. On Howes, see Christine von Oertzen, "Science in the Cradle: Milicent Shinn and Her Home-Based Network of Baby Observers, 1890–1910," *Centaurus* 55 (2013): 175–95.

28. On these networks, see Dorothy Ross, *G. Stanley Hall: The Psychologist as Prophet* (Chicago: University of Chicago Press, 1972), 287.

29. Von Oertzen, "Science in the Cradle," 178. See also Elissa N. Rodkey, "Far More Than Dutiful Daughter: Milicent Shinn's Child Study and Education Advocacy after 1898," *Journal of Genetic Psychology* 177, no. 6 (2016): 209–30.

30. Von Oertzen, "Science in the Cradle," 179. At the time, Shinn also worked as the managing editor of the *Overland Monthly*, a small but prestigious literary magazine.

31. Von Oertzen, "Science in the Cradle," 183.

32. Milicent Washburn Shinn, *The Biography of a Baby* (Boston: Houghton, Mifflin, 1900), 6.

33. Shinn, *Biography of a Baby*, 6, 7.

34. Shinn, *Biography of a Baby*, 7, 8. Not surprisingly, the neurological framework mostly dropped out of the picture for Shinn and her network of women observers, as it had for Emily Talbot. An exception is Laura Swain Tilley, who referred to it with respect to maturity of brain

centers: "The great step taken in uniting the visual and tactile and muscular perceptions seems to bring an advance all along the line. Physiologically speaking, the increased metabolism in several cerebral centers, and the enlarged network of functional communication between these, seems to stimulate circulation, to increase the tension on other paths of communication, and so to hasten their maturity." Laura Swain Tilley, "Record of the Development of Two Baby Boys," *Publications of the Association of Collegiate Alumnae*, series 3, no. 22 (June 1910): 30. Tilley also talks about mirrors; see below.

35. Milicent Washburn Shinn, *Notes on the Development of a Child*, vol. 2, *The Development of the Senses in the First Three Years of Childhood* (Berkeley: University of California Press, 1907), 4, 5.

36. Shinn talked about self-consciousness only in the sense of a difference between self and nonself, that is, consciousness of the child's own body and its boundaries, e.g., the section "Feeling of a Bodily Self," in *Notes on the Development of a Child*, 2:133–36.

37. Wilhelm Preyer, "Psychogenesis," trans. Marion Talbot, *Journal of Speculative Philosophy* 15, no. 2 (April 1881): 159–88.

38. Preyer to E. Talbot, November 22, 1880, in "The Development of Human Intelligence," *Nature* 23, no. 600 (April 28, 1881): 617–18, on 617–18.

39. Wilhelm Preyer, "Notes on the Development of Self-Consciousness: From *Die Seele des Kindes*," trans. Marion Talbot, *Education* 2 (1882): 290–99.

40. William Preyer, "A German Child," in E. Talbot, *Papers on Infant Development*, 44–48, on 44. Note that Marion Talbot's translation of Preyer's talk "Psychogenesis," the year before, did not touch on the *Ichgefühl* or mirror. Another shortened translation of "Psychogenesis" by W. H. Larrabee did not touch on the *Ichgefühl* or mirror either. William Preyer, "Psychogenesis in the Human Infant," trans. W. H. Larrabee, *Popular Science Monthly* 17 (1880): 625–35.

41. Note that these weren't entirely straightforward behaviors either. Especially the first "real" smile was the subject of debate.

42. E. Talbot, "Register of Infant Development," in E. Talbot, *Papers on Infant Development*, 51–52, on 51.

43. Emily Talbot, "American Children: Case A," in E. Talbot, *Papers on Infant Development*, 11–13, on 12.

44. E. Talbot, "Case A," 13 (my emphasis).

45. Tilley, "Record of the Development of Two Baby Boys," 9, 12, 40, 39, 82.

46. Tilley, "Record of the Development of Two Baby Boys," 82.

47. Tilley, "Record of the Development of Two Baby Boys," 82.

48. Robert N. Nye, "Medicine and Science as Masculine Fields of Honor," *Osiris* 12 (1997): 60–79. See also the more recent volume "Scientific Masculinities," ed. Erika Milam and Robert Nye, *Osiris* 30 (2015). On the gendering of American child study, see David Hoogland Noon, "Situating Gender and Professional Identity in American Child Study, 1880–1910," *History of Psychology* 7, no. 2 (2004): 107–29. On the relationship between cognitive and emotional approaches in medicine and education, see Marga Vicedo, *Intelligent Love: The Story of Clara Park, Her Autistic Daughter, and the Myth of the Refrigerator Mother* (Boston: Beacon, 2021).

49. E. Talbot, "Report," 5–6.

50. William T. Harris, "The Education of the Family, and the Education of the School," in E. Talbot, *Papers on Infant Development*, 1–5, on 5.

51. G. Stanley Hall, *Life and Confessions of a Psychologist* (New York: D. Appleton, 1923). On Hall, see D. Ross, *G. Stanley Hall*.

52. See, e.g., G. Stanley Hall, "Introduction to the American Edition," in *The Mind of the Child, Part I: The Senses and the Will; Observations concerning the Mental Development of the Human Being in the First Years of Life*, by Wilhelm Preyer, trans. H. W. Brown (New York: D. Appleton, 1888), xxi–xxv.

53. G. Stanley Hall, "Introduction to the American Edition," xxiii.

54. G. Stanley Hall, "Notes on the Study of Infants," *Pedagogical Seminary* 1, no. 2 (1891): 127–38. Hall's son was born on February 7, 1881, and his daughter on May 30, 1882. Hall published his diary at a time of personal crisis, just after the sudden death of his wife and eight-year-old daughter in an asphyxiation accident. As Hall's biographer Dorothy Ross suggested, he "fe[lt] perhaps that this portion of his life was now closed." D. Ross, *G. Stanley Hall*, 209.

55. Note, however, that Hall, in a discussion of the pathologies of speech, also criticized Preyer for going too far by including degenerative change in his recapitulation theory. Hall, "Notes on the Study of Infants," 133.

56. The Pädagogische Verein (Pedagogical Association) in Berlin, for example, had tested a total of twenty thousand children (ten thousand boys and ten thousand girls) from city and countryside, asking *Schulvorstände* (school boards) to conduct a questionnaire that tested "die Individualität der in die unterste Klasse eintretenden Schüler und Schülerinnen, so weit sie auf den Vorstellungen aus der Umgebung des Kindes beruht" (60), asking questions such as if they have seen a running hare, or a "im Freien hüpfender Frosch," or if they knew certain fairy tales. Education should be adjusted in response, e.g., through organizing excursions to the country-side for city students, using material objects in instruction (it was both assumed and confirmed in the study that city-based children were behind their peers from the countryside in knowledge of the world). F. Bartholomai and H. Schwabe, "Der Vorstellungskreis der Berliner Kinder beim Eintritt in die Schule," *Berlin und seine Entwicklung: Städtisches Jahrbuch für Volkswirtschaft und Statistik* 4 (1870): 59–77. Hall cited this work in 1893: *The Contents of Children's Minds on Entering School* (New York: E. L. Kellogg, 1893), 3. On the wider attempts to teach German children about nature, see Lynn Nyhart, *Modern Nature: The Rise of the Biological Perspective in Germany* (Chicago: University of Chicago Press, 2009), chap. 5. See also Sally Gregory Kohlstedt's book on the American nature study movement: *Teaching Children Science: Hands-On Nature Study in North America, 1890–1930* (Chicago: University of Chicago Press, 2010).

57. Kurt Danziger, "The Origins of the Psychological Experiment as a Social Institution," *American Psychologist* 40, no. 2 (1985): 133–40. See also Kurt Danziger, *Constructing the Subject: Historical Origins of Psychological Research* (Cambridge: Cambridge University Press, 1990).

58. Dr. G. Stanley Hall Collection, Archives and Special Collections, Robert H. Goddard Library, Clark University, Subseries 7: Topical Syllabi. B1-7-1 Topical Syllabi, 1894–1906, "The Early Sense of Self," 1895.

59. Even though "the early sense of self" questionnaire included unambiguous questions, it did broach a more complex and difficult topic, which is why, perhaps, it was only at the height of Hall's reputation inside and outside the psychological profession, in the mid-1890s, that he devised it.

60. G. Stanley Hall and John M. Mansfield, *Hints at a Selective and Descriptive Bibliography of Education* (Boston: D. C. Heath, 1886), 90.

61. In this way, Thorndike hoped to break free from the "anecdotal" method of evolutionists such as George Romanes, relying instead on experimentation and testing.

62. Edward L. Thorndike, *Animal Intelligence: An Experimental Study of the Associative Processes in Animals*, monograph supplement, *Psychological Review* 2, no. 4 (1898). Thorndike's choice of the title, the same as George Romanes's *Animal Intelligence* (London: K. Paul, Trench, 1882), for his dissertation might have been motivated by his wish to mark a methodological difference while focusing on the same subject.

63. John M. O'Donnell, *The Origins of Behaviorism: American Psychology, 1870–1920* (New York: New York University Press, 1985), 165–66.

64. Lewis M. Terman, "Trails to Psychology," in *A History of Psychology in Autobiography*, vol. 2, ed. Carl Murchison (Worcester, MA: Clark University Press, 1932), 297–31, on 318.

65. Terman, "Trails to Psychology," 318. Hall's response was this: "When I announced to him my decision he expressed very emphatically his disapproval of mental tests, but, finding that my mind was made up, he finally gave me his blessing and some advice on the danger of being misled by a quasi-exactness of quantitative methods." Terman, "Trails to Psychology," 318.

66. On Gesell, see Scott Curtis, "'Tangible as Tissue': Arnold Gesell, Infant Behavior, and Film Analysis," *Science in Context* 24, no. 3 (2011): 417–42; Carola Ossner, "Normal Development: The Photographic Dome and the Children of the Yale Psycho-clinic," *Isis* 111, no. 3 (2020): 515–41.

67. Arnold Gesell, "The Significance of the Nursery School: Excerpts Reprinted from *Child hood Education*, 1924, Vol. 1, No. 1," *Childhood Education* 93, no. 3 (2017): 194–98, on 196, 197. See also Arnold Gesell, "A Mental Hygiene Service for Pre-school Children," *American Journal of Public Health* 12, no. 12 (1922): 1030–33. On mental hygiene and the work of Adolf Meyer, see S. D. Lamb, *Pathologist of the Mind: Adolf Meyer and the Origins of American Psychiatry* (Baltimore: Johns Hopkins University Press, 2016).

68. In addition to Yale, these included institutes at Teachers College at Columbia University, the University of Minnesota, and the University of Toronto.

69. Rachel Stutsman, *Mental Measurement of Preschool Children: With a Guide for the Administration of the Merrill-Palmer Scale of Mental Tests* (Yonkers-on-Hudson, NY: World Book, 1931), 43.

70. Stutsman, *Mental Measurement of Preschool Children*, 4. See also Ellen Herman, "Families Made by Science: Arnold Gesell and the Technologies of Modern Child Adoption," *Isis* 92, no. 4 (December 2001): 684–715.

71. Ruth Griffiths, Rachel Stutsman, and Arnold Gesell did cross-sectional tests; Mary Shirley, Psyche Cattell, Charlotte Bühler, and Nancy Bayley did longitudinal tests. Only Shirley, however, made a point of the difference, and all, including Shirley, saw themselves as participating in the same research tradition. Ruth Griffiths, *The Abilities of Babies: A Study in Mental Measurement* (London: University of London Press, 1954); Stutsman, *Mental Measurement of Preschool Children*; Arnold Gesell and Helen Thompson, *Infant Behavior: Its Genesis and Growth* (New York: McGraw-Hill, 1934); Arnold Gesell and Catherine Amatruda, *Developmental Diagnosis: Normal and Abnormal Child Development* (New York: Paul B. Hoeber, 1941); Mary Shirley, *The First Two Years: A Study of Twenty-Five Babies*, 3 vols. (Minneapolis: University of Minnesota Press, 1931–33); Psyche Cattell, *The Measurement of Intelligence of Infants and Young Children* (1940; New York: Psychological Corporation, 1947); Charlotte Bühler, *The First Year of Life*,

trans. Pearl Greenberg and Rowena Ripin (New York: John Day, 1930); Nancy Bayley, *Mental Growth during the First Three Years: A Developmental Study of Sixty-One Children by Repeated Tests* (Worcester, MA: Clark University Press, 1933).

72. Griffiths, *Abilities of Babies*, 121. To give another example, for Nancy Bayley, "play to mirror" was one of a total of 185 test items. Cattell's testing equipment included a mirror and other items such as a key, a cup, a doll, a spoon, and a piece of string, Cattell, *Measurement of Intelligence*, 94.

73. All the books discussed here are from the 1920s and 1930s. Two exceptions are Psyche Cattell's from 1940 and Ruth Griffiths's from 1954. Gesell's writings also went into the 1940s.

74. Stutsman, *Mental Measurement of Preschool Children*, chap. 3.

75. Griffiths, *Abilities of Babies*, 127.

76. Gesell and Thompson, *Infant Behavior*, 240.

77. Gesell and Thompson, *Infant Behavior*, 106, 154, 239.

78. Arnold Gesell and Louise Ames, "The Infant's Reaction to His Mirror Image," *Pedagogical Seminary and Journal of Genetic Psychology* 70, no. 2 (1947): 141–54. In *Infant Behavior*, Gesell and Thompson list eleven mirror-behavior items: sober [response], smiles, vocalizes, waves arms, brings hand to mirror, pats mirror, approaches image socially, brings face to mirror, plays peekaboo with image, postural activity, and stands (241).

79. Rather, the behaviors listed showed a trend toward increasing social interaction with the mirror image. The recordings go only up to sixty weeks, which confirms Gesell's lack of interest in self-recognition.

80. Gesell was not the only one to use film during this time. A range of psychologists including John B. Watson, Kurt Lewin, and Myrtle McGraw made use of it as a research tool, thus contributing to the professionalization of developmental psychology. Curtis, "'Tangible as Tissue.'" Film was an expensive technology at the time, and, as Carola Ossner has shown, the need to save materials led Gesell to the selective choice of "only the most uniform and typical examination situations," thus reinforcing the selective normativity of Gesell's project. Ossner, "Normal Development," 539. There is no indication that Gesell did this more fine-grained film-based analysis for other behaviors discussed in earlier books.

81. Arnold Gesell, "Cinemanalysis: A Method of Behavior Study," *Pedagogical Seminary and Journal of Genetic Psychology* 47, no. 1 (1935): 3–16, on 7.

82. Gesell, "Cinemanalysis," 5.

83. Gesell and Thompson, *Infant Behavior*, 8.

84. Gesell and Thompson, *Infant Behavior*, 11.

85. Bühler was referring here to John B. Watson's behaviorism. This is interesting, especially because the work of the gestalt psychologists suggested that there were no single reflexes and that all behavior was holistic.

86. Bühler, *First Year of Life*, 13.

87. Bühler, *First Year of Life*, 14.

88. Bühler, *First Year of Life*, all quotes on 242. Bühler did other tests at later moments, but they did not differ in principle. Note the normative character of this approach. See also Alice Smuts, *Science in the Service of Children, 1893–1935* (New Haven, CT: Yale University Press, 2006).

89. Bühler, *First Year of Life*, 14.

90. Cattell, *Measurement of Intelligence*, 85.

91. Bayley, *Mental Growth during the First Three Years*, 27, 31.

92. Shirley, *First Two Years*, 2:263–65.

93. Gesell and Thompson, *Infant Behavior*, e.g., 161; Griffiths, *Abilities of Babies*, e.g., 159.

94. Bühler, *First Year of Life*, e.g., 240.

95. Bühler, *First Year of Life*, 194.

96. Bühler, *First Year of Life*, 200.

97. Shirley, *First Two Years*, 2:262, 264.

98. Stutsman, *Mental Measurement of Preschool Children*, 170.

99. Stutsman, *Mental Measurement of Preschool Children*, 171. Another example is in Griffiths, *Abilities of Babies*, 121.

100. Stutsman, *Mental Measurement of Preschool Children*, 170–71.

101. Griffiths, *Abilities of Babies*, 14.

102. John B. Watson, "Psychology as the Behaviorist Views It," *Psychological Review* 20, no. 2 (1913): 158–77, on 158.

103. On the Wundtian tradition, see Danziger, *Constructing the Subject*.

104. Of course experiments on humans and in particular infants were conducted as well, with Watson's on Little Albert being only the most notorious one. On Watson and Little Albert, see Ben Harris, "Whatever Happened to Little Albert?," *American Psychologist* 34, no. 2 (1979): 151–60.

105. John B. Watson, *Psychology from the Standpoint of a Behaviorist* (Philadelphia: J. B. Lippincott, 1919), 10. A stimulus, for Watson, could be external, that is, referring to an environmental situation, but also internal, referring to an inner condition of the organism. Note, however, that Watson radicalized over time.

106. John B. Watson, *Behaviorism* (1924; New Brunswick: Transaction, 2009), 165.

107. Watson performed these experiments in collaboration with Karl Lashley. John B. Watson, "The Place of the Conditioned Reflex in Psychology," *Psychological Review* 23 (1916): 89–116, on 96–97.

108. John B. Watson and Rosalie Rayner, "Conditioned Emotional Reactions," *Journal of Experimental Psychology* 3, no. 1 (1920): 1–14.

109. Baldwin was caught in a Baltimore brothel during a police raid. Wade E. Pickren and Alexandra Rutherford, *A History of Modern Psychology in Context* (New York: John Wiley and Sons, 2010), 60.

110. Arnold Gesell, "The Conditioned Reflex and the Psychiatry of Infancy," *American Journal of Orthopsychiatry* 8, no. 1 (1938): 19–30, on 28.

111. Gesell, "Conditioned Reflex," 19.

112. The conditions were: position of the baby (horizontal or vertical), nature of stimulus (the mirror, a parent, the experimenter), and the simultaneous view of two faces (a familiar face or an unfamiliar face). Geneviève Balleyguier, "Premières réactions devant le miroir," *Enfance* 17, no. 1 (1964): 51–67.

113. Allan H. Schulman and Cheryl Kaplowitz, "Mirror-Image Response during the First Two Years of Life," *Developmental Psychobiology* 10, no. 3 (1977): 133–42, on 135.

114. Schulman and Kaplowitz noted thirteen different behaviors, thirty-four if considering the different mirror conditions. Despite these complexities, they discovered stage-like patterns in the infants' behaviors.

115. The publication was based on parts of her dissertation: Ann Bigelow, "The Correspondence between Self- and Image Movement as a Cue to Self-Recognition for Young Children," *Journal of Genetic Psychology* 139, no. 1 (1981): 11–26.

116. For details, see Bigelow, "Correspondence between Self- and Image Movement."

117. Hanuš Papoušek and Mechthild Papoušek, "Mirror Image and Self-Recognition in Young Human Infants: I. A New Method of Experimental Analysis," *Developmental Psychobiology* 7, no. 2 (1974): 149–57. This was confirmed by Beulah Amsterdam and Lawrence M. Greenberg, "Self-Conscious Behavior of Infants: A Videotape Study," *Developmental Psychobiology* 10, no. 1 (1977): 1–6.

118. Bennett I. Bertenthal and Kurt W. Fischer, "Development of Self-Recognition in the Infant," *Developmental Psychology* 14, no. 1 (1978): 44–50, on 49.

119. J. C. Dixon, "Development of Self Recognition," *Journal of Genetic Psychology* 91, no. 2 (1957): 251–56. In this study, Dixon also reported his and Dorothy Dixon's studies with twins where one child saw its twin through the glass, and the other saw itself in the mirror. See the interlude for René Zazzo's and Anne-Marie Fontaine's development of these twin mirror experiments. Dorothy Dixon, "The Mirror Behavior of Twins" (MA thesis, University of Florida, 1952).

120. The cutesy name is misleading because of course at that age children are not speaking yet.

121. Beulah Amsterdam, "Mirror Self-Image Reactions before Age Two," *Developmental Psychobiology* 5, no. 4 (1972): 297–305, on 302.

122. Amsterdam, "Mirror Self-Image Reactions," 304.

123. Beulah Amsterdam, "Mirror Behavior in Children under Two Years of Age" (PhD diss., University of North Carolina at Chapel Hill, 1968), 18.

124. Amsterdam, "Mirror Behavior," 20.

125. Note that the test for Amsterdam was never simple. From the beginning, she believed that mirror self-recognition had to be considered as part of a behavioral sequence. In a pilot study for her 1968 PhD, Amsterdam had already worked out that there was a developmental sequence in the child leading up to self-recognition; in that study, self-recognition was tested verbally, which the rouge test was meant to replace. Amsterdam, "Mirror Behavior."

Chapter 3: The Dancing Robot

1. William Grey Walter, "Presentation: Dr. Grey Walter," in *Discussions on Child Development*, ed. J. M. Tanner and Bärbel Inhelder, vol. 2 (1956; New York: International Universities Press, 1971), 21–74, on 36.

2. William Grey Walter, *The Living Brain* (New York: Norton, 1953), 128.

3. Walter, "Presentation," 36.

4. Walter, *Living Brain*, 130.

5. "Bristol's Robot Tortoises Have Minds of Their Own," *BBC Newsreel*, February 17, 1950.

6. They were from the batch of tortoises built by the Burden Neurological Institute (BNI) engineer W. J. "Bunny" Warren in early 1951. Owen Holland, "Exploration and High Adventure: The Legacy of Grey Walter," *Philosophical Transactions of the Royal Society A* 361, no. 1811 (October 15, 2003): 2085–121, on 2092; Andrew Pickering, *The Cybernetic Brain: Sketches of Another Future* (Chicago: University of Chicago Press, 2010), 53. Note that Elsie the Cow and Elmer the Bull were the names of advertising animals for the Borden Dairy Company that had been

created in the 1930s. It is not clear if Walter knew of the US company or, if he did, how their status as cows related to his animal hierarchy.

7. Archival documents posted online, accessed June 2016: ES106 "Machina Speculatrix": (Mechanical Tortoises), Dr. Grey Walter, Burden Neurological Institute, Stapleton, Bristol, Archival record: FOB/4922; National Archives at Kew, Richmond, Surrey, South Bank Exhibition, Festival of Britain, 1951. The two tortoises—mechanical devices actuated by a two-cell electronic brain—were exhibited in the "How We Know" section in September 1951. On the Festival of Britain, see Mary Banham and Bevis Hillier, eds., *A Tonic to the Nation: The Festival of Britain 1951* (London: Thames and Hudson, 1976); Becky E. Conekin, *"The Autobiography of a Nation": The 1951 Festival of Britain* (Manchester: Manchester University Press, 2003).

8. B. A. Young, "Sorry, No Miracles: The Exhibition of Science, South Kensington," *Punch* (London), June 6, 1951, 682–83, on 683.

9. It is currently not on view. A successor of the original tortoises (Walter's "#6") was also on display at the London Science Museum but moved to the basement in 2009. There, it was examined in great detail and photographed by the engineer and computer scientist David Buckley (along with Reuben Hoggett), who posted a gallery of photographs on the internet at davidbuckley.net, "History Making Robots."

10. Rhodri Hayward, "The Tortoise and the Love-Machine: Grey Walter and the Politics of Electroencephalography," *Science in Context* 14, no. 4 (2001): 615–41, on 616.

11. Owen Holland, "Grey Walter: The Pioneer of Real Artificial Life," in *Artificial Life V. Proceedings of the Fifth International Workshop on the Synthesis and Simulation of Life*, ed. Christopher G. Langton and Katsunori Shimohara (Cambridge, MA: MIT Press, 1996), 33–41.

12. Pickering, *Cybernetic Brain*, chap. 3; Rhodri Hayward, "'Our Friends Electric': Mechanical Models of Mind in Postwar Britain," in *Psychology in Britain: Historical Essays and Personal Reflections*, ed. G. C. Bunn, A. D. Lovie, and G. D. Richards (Leicester: British Psychological Society, 2001), 290–308; Hayward, "Tortoise and the Love-Machine"; Cornelius Borck, *Hirnströme: Eine Kulturgeschichte der Elektroenzephalographie* (Göttingen: Wallstein, 2005), esp. 301–12; Cornelius Borck, "Vital Brains: On the Entanglement of Media, Minds, and Models," *Progress in Brain Research* 233 (2017): 1–23.

13. Pickering, *Cybernetic Brain*, 48–49. For an account of the relationship between machines and the higher function of love, see Hayward, "Tortoise and the Love-Machine."

14. Norbert Wiener, *Cybernetics; or, Control and Communication in the Animal and the Machine*, 2nd ed. (1948; Cambridge, MA: MIT Press, 1961). The same is true for historians of cybernetics. As Pickering has suggested, "everyone can have their own history of cybernetics." Pickering, *Cybernetic Brain*, 3. Scholars have highlighted the multifaceted nature of cybernetics, from its links with US military research (Peter Galison) through its international aspects (Eden Medina) to the importance of the brain, especially in the British tradition (Borck, Pickering, Hayward, Tara Abraham). Peter Galison, "The Ontology of the Enemy: Norbert Wiener and the Cybernetic Vision," *Critical Inquiry* 21 (1994): 228–66; Eden Medina, *Cybernetic Revolutionaries: Technology and Politics in Allende's Chile* (Cambridge, MA: MIT Press, 2011); Borck, *Hirnströme*; Pickering, *Cybernetic Brain*; Hayward, "Tortoise and the Love-Machine"; Hayward, "'Our Friends Electric'"; Tara Abraham, *Rebel Genius: Warren S. McCulloch's Transdisciplinary Life in Science* (Cambridge, MA: MIT Press, 2016). See also Ronald Kline, *The Cybernetics Moment; or, Why We Call Our Age the Information Age* (Baltimore: Johns Hopkins University Press, 2015).

My chapter builds on this work. For an excellent overview of the literature, see Abraham, *Rebel Genius*, introduction.

15. Wiener, *Cybernetics*, 6–7, 7.

16. Papers of the Ratio Club, GC-179-B, Wellcome Trust Archives and Manuscripts, London, Announcement of talk to be given by Walter and H. W. Shipton at the Burden Neurological Institute, May 7, 1955: "Discussion on control mechanisms in machines, animals, and communities." The full quote is: "The first attempt to unify these subjects is the book by Wiener, who has coined the name 'CYBERNETICS' to describe all studies of communication, semantics, control, calculation, adaptation, behavior, teleology, evolution, stochastics and economics, whatever the nominal branch of science within which they may be originated."

17. Letter, Walter to John Bates, September 29, 1949; Bates to Walter, October 4, 1949, Papers of the Ratio Club, GC-179-B2 and GC-179-B3.

18. Letter, Bates to Walter, July 27, 1949, Papers of the Ratio Club, GC-179-B1.

19. For a history of the Ratio Club, see Philip Husbands and Owen Holland, "The Ratio Club: A Hub of British Cybernetics," in *The Mechanical Mind in History*, ed. Philip Husbands, Owen Holland, and Michael Wheeler (Cambridge, MA: MIT Press, 2008), 91–148. See also Margaret Boden, "Grey Walter's Anticipatory Tortoises," *Rutherford Journal* 2 (2006–7): http://www.rutherfordjournal.org/article020101.html.

20. On the former, it would be going too far to talk of a national style here. Even if in Wiener's work the relation with neuroscience was less prominent, the ties between the neurodisciplines and cybernetics were strong in the United States as well. See Abraham, *Rebel Genius*, 12–13.

21. Letter, Bates to Walter, July 27, 1949, Papers of the Ratio Club, GC-179-B1. For a full list of members, see Husbands and Holland, "Ratio Club," 94–98.

22. Husbands and Holland, "Ratio Club," 91.

23. On the trope of mechanizing the mind, see Hayward, "'Our Friends Electric'"; Husbands, Holland, and Wheeler, *Mechanical Mind in History*; Margaret Boden, *Mind as Machine: A History of Cognitive Science* (Oxford: Oxford University Press, 2006).

24. Papers of the Ratio Club, GC-179-B5.

25. Papers of the Ratio Club, GC-179-B5.

26. Papers of the Ratio Club, GC-179-B5.

27. W. Grey Walter, "Neurocybernetics (Communication and Control in the Living Brain)," in *Survey of Cybernetics: A Tribute to Dr. Norbert Wiener*, ed. J. Rose (London: Iliffe Books, 1969), 93–108, on 93–94.

28. W. Grey Walter, "An Improved Low Frequency Analyser," *Electronic Engineering* 16 (1943): 236–38.

29. Walter, "Neurocybernetics," 94.

30. Walter, *Living Brain*, chap. 5; W. Grey Walter, "An Imitation of Life," *Scientific American* 182, no. 5 (May 1950): 42–45; W. Grey Walter, "A Machine That Learns," *Scientific American* 185, no. 2 (August 1951): 60–63. Note that *The Living Brain* is a problematic source, which, according to Grey Walter's son Nicolas Walter, had been written by his father "from his notes and conversations." Quoted in Holland, "Exploration and High Adventure," 2088. It is also not clear to what extent his wife, Vivian Walter née Dovey, a radiologist at the Burden, as well as his technician Bunny Warren helped in the construction of the machine. Dovey coauthored several papers with Walter, but her memory is mostly erased in the archives of the Burden. See David Saunders,

"Wired-Up in White Organdie: Framing Women's Scientific Labour at the Burden Neurological Institute," *Science Museum Group Journal*, no. 10 (2018): https://dx.doi.org/10.15180%2F181003.

31. Walter also played an important role in the institutionalization of EEG research: He founded the EEG Society in 1943, organized the first EEG Congress in 1947, founded the International Federation of EEG Societies, and was its president from 1953 to 1957. He also founded the *EEG Journal*. Holland, "Exploration and High Adventure," 2087. On W. Grey Walter, see also John Johnston, *The Allure of Machinic Life: Cybernetics, Artificial Life, and the New AI* (Cambridge, MA: MIT Press, 2008), 47–53.

32. Berger's first paper was Hans Berger, "Über das Elektrenkephalogramm des Menschen," *Archiv für Psychiatrie und Nervenheilkunde* 87 (1929): 527–70. For a reconstruction of the early history of EEG research, see Borck, "Hans Bergers langer Weg zum EEG," in *Hirnströme*, 23–84.

33. W. Grey Walter, "Thought and Brain: A Cambridge Experiment," *Spectator* (London), October 5, 1934, 478–79, on 479.

34. E. D. Adrian and B.H.C. Matthews, "The Interpretation of Potential Waves in the Cortex," *Journal of Physiology* 81, no. 4 (1934): 440–71.

35. Jan-Friedrich Tönnies, "Die unipolare Ableitung elektrischer Spannungen vom menschlichen Gehirn," *Naturwissenschaften* 22 (1934): 411–14.

36. E. D. Adrian and B.H.C. Matthews, "The Berger Rhythm: Potential Changes from the Occipital Lobes in Man," *Brain* 57 (1934): 355–85.

37. Adrian and Matthews, "Berger Rhythm," 360; W. Grey Walter, "The Electroencephalogram in Cases of Cerebral Tumour," *Journal of the Royal Society of Medicine* 30, no. 5 (1937): 579–98, on 583.

38. Walter, *Living Brain*, 84.

39. Adrian and Matthews, "Berger Rhythm," 382.

40. W. Grey Walter "The Location of Cerebral Tumours by Electro-encephalography," *Lancet* 228 (August 8, 1936): 305–8.

41. Walter, *Living Brain*, 84.

42. Walter, "Electro-encephalogram in Cases of Cerebral Tumour," 583.

43. W. Grey Walter, "Electroencephalography," in *Annual Report of the Board of Regents of the Smithsonian Institution . . . 1950*, H.R. Doc. 9/1 (Washington, DC, 1951), 243–53, on 244.

44. Quoted in Pickering, *Cybernetic Brain*, 409n23. This was the age of somatic treatments in psychiatry, and Walter's home institution was at the forefront of these developments in Britain. For instance, it saw the first prefrontal leucotomy in Britain in 1940. Ray Cooper, "Research at the Burden Neurological Institute, Bristol," *Bio-medical Engineering* 7, no. 5 (1972): 220–25, on 221. See also R. Cooper and J. Bird, *The Burden: Fifty Years of Clinical and Experimental Neuroscience at the Burden Neurological Institute* (Bristol: White Tree Books, 1989).

45. On Craik and Walter, see Hayward, "'Our Friends Electric,'" 295–99.

46. Kenneth J. W. Craik, "Theory of the Human Operator in Control Systems: I. The Operator as an Engineering System," *British Journal of Psychology* 38, no. 2 (1947): 56–61, on 56. See also Hayward, "'Our Friends Electric,'" 296.

47. W. Grey Walter, "Features in the Electrophysiology of Mental Mechanisms," in *Perspectives in Neuropsychiatry: Essays Presented to Professor Frederick Lucien Golla by Past Pupils and Associates* (London: H. K. Lewis, 1950), 67–78, on 69. Scanning was one of the key principles

of cybernetics, often in self-conscious inspiration from wartime technology. Borck, *Hirnströme*, 298–301; Pickering, *Cybernetic Brain*, 45.

48. Walter, *Living Brain*, 104.

49. Walter, *Living Brain*, 107–8.

50. Walter, *Living Brain*, 71.

51. Walter, *Living Brain*, 108.

52. Walter, "Features in the Electrophysiology of Mental Mechanisms," 74.

53. Walter, *Living Brain*, 101.

54. Walter, "Features in the Electrophysiology of Mental Mechanisms," 74.

55. Walter, "Features in the Electrophysiology of Mental Mechanisms," 74.

56. Papers of the Ratio Club, Talk by Walter, "Pattern Recognition," May 15, 1950, GC-179-B6.

57. Walter, "Features in the Electrophysiology of Mental Mechanisms," 71.

58. Walter, "Presentation," 31.

59. Walter, "Features in the Electrophysiology of Mental Mechanisms," 71.

60. Other science sections included Plant Morphogenesis, Cancer, and the Problem of Life. *1951 Exhibition of Science, South Kensington, Festival of Britain, Guide-Catalogue*, 50, Dr. W. Grey Walter, Advice and supervision of Model of Electrical Tortoises, WORK 25/257/G1/C2/647, 1948–51, National Archives, Kew.

61. *1951 Exhibition of Science, South Kensington, Festival of Britain, Guide-Catalogue*, 32, Dr. W. Grey Walter, Advice and supervision of Model of Electrical Tortoises, WORK 25/257/G1 /C2/647, 1948–51, National Archives, Kew.

62. Walter, "Imitation of Life," 43.

63. Walter, *Living Brain*, 287.

64. Quoted in Walter, "Imitation of Life," 42.

65. Walter, *Living Brain*, 130.

66. Walter, "Presentation," 26.

67. This diverts from Jessica Riskin's reading: "All of the Tortoises' defining features— restlessness, exploratory curiosity, self-recognition, sociability—were believable only when one disregarded the internal mechanisms that caused these appearances. One had to equate appearance with reality, to decide that something seemingly conscious, curious, restless, or sociable, was indeed conscious, curious, restless, or sociable." Jessica Riskin, *The Restless Clock: A History of the Centuries-Long Argument over What Makes Living Things Tick* (Chicago: University of Chicago Press, 2016), 324. Walter didn't, however, disregard the internal mechanisms—quite the opposite. Rather he wanted to deploy his understanding of the robots' internal mechanisms to give new accounts of consciousness, curiosity, etc.

68. He continued: "In both cases the function is primarily one of economy, just as in a television system the scanning of the image permits transmission of hundreds of thousands of point-details on one channel instead of as many channels." Walter, "Imitation of Life," 44.

69. Walter, "Imitation of Life," 43.

70. This is unlike Alice when the White Queen quizzed her: "Can you do addition? . . . What's one and one and one and one and one and one and one and one and one and one?" Quoted in Walter, *Living Brain*, 121.

71. Walter, "Imitation of Life," 43.

72. Walter, *Living Brain*, 131.

73. Walter, "Imitation of Life," 43.

74. Walter, *Living Brain*, 131. See below for an explanation of Walter's pervasive mirror metaphors.

75. Walter, "Imitation of Life," 44.

76. Walter, "Presentation," 28.

77. This section heading is the title of the first chapter in *The Living Brain*.

78. Walter, *Living Brain*, 15–16.

79. Walter, *Living Brain*, 38.

80. Walter, *Living Brain*, 141, 142.

81. Walter, *Living Brain*, 142.

82. Walter, *Living Brain*, 139.

83. Walter, *Living Brain*, 149.

84. Walter, *Living Brain*, 148.

85. Walter, *Living Brain*, 16–17.

86. Walter, *Living Brain*, 16.

87. Walter, *Living Brain*, 38.

88. Walter, *Living Brain*, 16.

89. On homeostasis in cybernetics, see Riskin, *Restless Clock*, chap. 9.

90. Walter, *Living Brain*, 36.

91. Walter, *Living Brain*, 37.

92. Quoted in Walter, *Living Brain*, 36 (ellipses and emphasis in source).

93. On the difference between real and imaginary events, see W. Grey Walter, "The Twenty-Fourth Maudsley Lecture: The Functions of Electrical Rhythms in the Brain," *Journal of Mental Science* 96 (January 1950): 1–31, on 9.

94. See, e.g., W. Grey Walter, "Patterns in Your Head," *Discovery* (February 1952): 56–62.

95. Walter, *Living Brain*, 137.

96. Walter, *Living Brain*, 151. For the definite biography on Pavlov, see Daniel P. Todes, *Ivan Pavlov: A Russian Life in Science* (Oxford: Oxford University Press, 2014). Todes suggests that "conditional," as I've used here, is a better translation than "conditioned"; I've retained "conditioned" elsewhere as this is the term used by the historical actors.

97. Walter, "Patterns in Your Head," 57.

98. Walter, *Living Brain*, 123. That Walter compared it to a sleeping cat or dog does not imply that the homeostat could be compared to the evolutionary development of mammals. Rather, as Walter pointed out, the *M. sopora* would be classified by the naturalist "as a plant." Walter, *Living Brain*, 124.

99. Walter, *Living Brain*, 179.

100. It is worth noting that the mirror and mirror self-recognition also emerged as metaphors for his research. In his foreword to *The Living Brain*, Walter noted that "one or two listeners" of the BBC Home Service (the predecessor BBC Radio 4), which had featured Walter's work, "felt a kind of impudicity about brain surveying brain, as if suddenly coming upon themselves for the first time naked in a looking-glass" (11). Not only was his research a form of mirroring; it proceeded through various forms of mirroring. Walter understood his robots as "mirrors" for the brain, in the sense that they mimicked the brain's functions, while the EEG was also a

"mirror," externalizing the mind so it could examine itself. As Walter wrote, "EEG records may be considered, then, as the bits and pieces of a mirror for the brain, itself *speculum speculorum*" (60). The pervasiveness of mirror metaphors in Walter's book was reinforced by a running reference to Lewis Carroll's *Alice through the Looking-Glass*. The epigraph of chapter 2, "A Mirror for the Brain" (where the introductory account of brain evolution of the previous pages was set to one side in favor of a more detailed functional analysis of its workings), was taken from the opening scenes of Carroll's novel, just before Alice enters the world on the other side of the looking glass: "Let's pretend there's a way of getting through it somehow, Kitty" (quoted on 40).

101. Walter, *Living Brain*, 128.

102. Archives of the Burden Neurological Institute, Science Museum Library and Archives, Science Museum at Wroughton, A-6–32 (emphasis in original).

103. Walter, *Living Brain*, 130.

104. Later researchers nevertheless took the joke seriously. Junichi Takeno, a robotics engineer at Meiji University in Tokyo, in a book published in 2013, wrote that Walter had invented the "world's first life-simulating robot," and he placed great emphasis on the mirror test. Considering it "a wonder that such an experiment was successfully conducted in as early as 1950," Takeno argued that it helped unravel "the mystery of human consciousness." Conventional AI, he thought, was "at a deadlock," and following Walter's lead, Takeno sought to "construct a totally new AI," what he called "artificial consciousness (AC)." The proof of this consciousness came when Takeno placed his own robot in front of the mirror, which he compared to an encounter with another robot that was wired to and controlled by it, and to an encounter with another, independent robot. The robot was able to distinguish between the three cases by the "coincidence rate" between the performed and observed movements, and thus was able to assess whether it was dealing with an image of itself, or another machine. That is why Takeno thought he could talk about the robot "cognizing itself in the mirror." Junichi Takeno, *Creation of a Conscious Robot: Mirror Image Cognition and Self-Awareness* (Singapore: Pan Stanford, 2013), 128, 131, 154. On the larger project of "understanding life by reproducing it," see Jessica Riskin, ed., *Genesis Redux: Essays in the History and Philosophy of Artificial Life* (Chicago: University of Chicago Press, 2007), 1, 2.

105. Note that Walter, although he did not suggest that the robot recognized itself, did believe that a well-defined cybernetic mechanism could be the basis of self-recognition. He chose the term "reflexive" to describe the tortoise's behavior in front of the mirror. J. M. Tanner and Bärbel Inhelder, eds., *Discussions on Child Development*, vol. 1 (1956; New York: International Universities Press, 1971), 177.

Chapter 4: Monkeys, Mirrors, and Me

1. The description is taken from Gordon G. Gallup Jr., "Mirror-Image Stimulation," *Psychological Bulletin* 70, no. 6 (1968): 782–93, on 788–89. This publication was based on Gallup's PhD dissertation research.

2. Gordon G. Gallup Jr., "Mirror-Image Stimulation and Psychological Research" (PhD diss., Washington State University, 1968), 46.

3. Gordon G. Gallup Jr., "Chimpanzees: Self-Recognition," *Science* 167 (1970): 86–87, on 87.

4. Frederick Rudolph, *Curriculum: A History of the American Undergraduate Source of Study since 1636* (San Francisco: Jossey-Bass, 1977), 117. As a result of the Morrill Acts of 1862 and 1890,

land was made available to the states, who could sell it to found educational institutions. As more recent work has emphasized, the land was stolen from indigenous people. Tristan Ahtone and Robert Lee, "Ask Who Paid for America's Universities," *New York Times*, May 7, 2020, https://www.nytimes.com/2020/05/07/opinion/land-grant-universities-native-americans.html. Some land grant universities, such as Cornell, have begun to work through this aspect of their past.

5. 1958 report on Education building, UA 333 WSU News Subject Files, box 9, folder 26, Washington State University Archives.

6. Dean T. H. Kennedy met with Elder in Tennessee, where Elder worked at the time, for the negotiations. Oral Histories, Archives 202, WSU Centennial Oral Histories, James H. Elder, Interview conducted January 19, 1989, Washington State University Archives.

7. Oral Histories, Archives 202, WSU Centennial Oral Histories, James H. Elder, Washington State University Archives.

8. Mary Kientlze, Psychology at Washington State University: A Brief History, January 1976, 5, Washington State University Archives, Digital Collections, Departmental Histories, http://content.libraries.wsu.edu/cdm/ref/collection/wsu_histor/id/3395.

9. UA 33, WSU News, box 28, folder 9, Washington State University Archives.

10. UA 333, WSU News Subject Files, box 12, folder 43, Washington State University Archives.

11. "Psychology at Washington State, 1976," Washington State University Archives, 5.

12. UA 333, WSU News Subject Files, box 28, folder 9 Washington State University Archives. Between the demolition of the old post office in 1964 and the completion of the new building in 1966, Klopfer's primates were housed in the university's maintenance building.

13. Francis Young received a $220,000 building grant from the NSF for the Comparative Behavior Laboratory. UA 333, WSU News Subject Files, box 28, folder 9, Washington State University Archives.

14. UA 333, WSU News Subject Files, box 11, folder 1, Washington State University Archives.

15. UA 333, WSU News Subject Files, box 8, folder 58, Washington State University Archives.

16. The faculty grew from eight in 1949–50 to twenty-three in 1967–68. "Psychology at Washington State, 1976," Washington State University Archives, 6.

17. Oral Histories, Archives 202, WSU Centennial Oral Histories, James H. Elder, Washington State University Archives.

18. Gallup, "Mirror-Image Stimulation and Psychological Research," 29; Gallup, "Mirror-Image Stimulation," 787.

19. Gallup, "Mirror-Image Stimulation and Psychological Research," 30; Gallup, "Mirror-Image Stimulation," 787.

20. Gallup, "Mirror-Image Stimulation," 787. As Gallup pointed out, mirrors were also used in social facilitation. For example, they could replace a conspecific to induce feeding responses.

21. Gallup, "Mirror-Image Stimulation and Psychological Research," 4.

22. See the WSU course catalogue of 1964–66, in which psychology was presented as "the scientific study of human and animal behavior." As further specified in the catalogue, "through the study of this field the student becomes acquainted with the systematic nature of human behavior, with techniques for investigating it, and with theories of this development." The catalogue of 1964–65 and 1965–66 was the first to include such a definition of psychology. WSU Publications 4, Catalogue, Washington State University Archives.

23. Nadine Weidman, *Constructing Scientific Psychology: Karl Lashley's Mind-Brain Debates* (Cambridge: Cambridge University Press, 1999), 12; John M. O'Donnell, *The Origins of Behaviorism: American Psychology, 1870–1920* (New York: New York University Press, 1985).

24. James H. Elder, "Robert M. Yerkes and Memories of Early Days in the Laboratories," in *Progress in Ape Research*, ed. Geoffrey Bourne (New York: Academic, 1977), 29–38, on 29. On the behaviorism of Theodor Beer, Albrecht Bethe, and Jakob von Uexküll, see Florian Mildenberger, "The Beer/Bethe/Uexküll Paper (1899) and Misinterpretations Surrounding 'Vitalistic Behaviorism,'" *History and Philosophy and the Life Science* 28, no. 2 (2006): 175–89. On Lloyd Morgan and Jacques Loeb, see Rebecca Lemov, *World as Laboratory: Experiments with Mice, Mazes, and Men* (New York: Hill and Wang, 2005). On John B. Watson, see Kerry Buckley, *Mechanical Man: John Broadus Watson and the Beginnings of Behaviorism* (New York: Guilford, 1989).

25. Gallup was continuously employed by the WSU Department of Psychology as a graduate student (master's and PhD), first as research assistant (1963–65), then as teaching assistant (1965–67), and then as instructor in psychology (1967–68). Washington State University Archives, MS 198726, box 18.

26. Donald A. Dewsbury, *Monkey Farm: A History of the Yerkes Laboratories of Primate Biology, Orange Park, Florida, 1930–1965* (Lewisburg, PA: Bucknell University Press, 2006). On Yerkes at Yale, see also Donna Haraway, *Primate Visions: Gender, Race, and Nature in the World of Modern Science* (New York: Routledge, 1989), 59–83.

27. Quoted in Elder, "Robert M. Yerkes and Memories of Early Days in the Laboratories," 32.

28. Elder, "Robert M. Yerkes and Memories of Early Days in the Laboratories," 33.

29. Elder, "Robert M. Yerkes and Memories of Early Days in the Laboratories," 31.

30. Noam Chomsky, review of *Verbal Behavior*, by B. F. Skinner, *Language* 35 (1959): 26–57. The review has been presented as misleading, e.g., in Marc N. Richelle, *B. F. Skinner: A Reappraisal* (Hove, UK: L. Erlbaum); William O'Donohue and Kyle E. Ferguson, *The Psychology of B. F. Skinner* (Thousand Oaks, CA: Sage, 2001).

31. P. W. Bridgman, *The Logic of Modern Physics* (New York: Macmillan, 1927), 5.

32. The same example was also suited to illustrating the complications of operationalizing it; see Joel Isaac, *Working Knowledge: Making the Human Sciences from Parsons to Kuhn* (Cambridge, MA: Harvard University Press, 2012), 102–7.

33. Historians have explained this enthusiasm in different ways, from the search for legitimation of the field to the idiosyncrasies of the "Harvard interstitial academy." For an account of this history, see Isaac, *Working Knowledge*.

34. He used the term "internal stimulus" in John B. Watson, *Behaviorism* (1924; New Brunswick: Transaction, 2009), 103.

35. Watson, *Behaviorism*, 6.

36. S. S. Stevens, "The Operational Basis of Psychology," *American Journal of Psychology* 47, no. 2 (1935): 323–30, on 330.

37. John A. Mills, *Control: A History of Behavioral Psychology* (New York: New York University Press, 1998), 183. It is quite possible that Gallup used the textbook when he was an undergraduate at WSU between 1959 and 1963.

38. Benton J. Underwood, *Experimental Psychology*, 2nd ed. (New York: Appleton-Century-Crofts, 1966), 299.

39. Underwood, *Experimental Psychology*, 300. Note that this is a change from the first edition of his textbook: Brenton J. Underwood, *Experimental Psychology: An Introduction* (New York: Appleton-Century-Crofts, 1949).

40. A symposium held in Boston in 1953 at the annual meeting of the American Association for the Advancement of Science agreed on this point, with Vienna Circle philosophers such as Gustav Bergmann and Carl Hempel raising the most critical voices. Isaac, *Working Knowledge*, 112.

41. This was at least the genealogy proposed by the historical actors. See, e.g., Stevens, "Operational Basis of Psychology," 323; Edward Tolman, "An Operational Analysis of 'Demands,'" *Erkenntnis* 6 (1936): 383–92, on 390. Historians have since offered more nuanced accounts of their affinities; see esp. Laurence D. Smith, *Behaviorism and Logical Positivism: A Reassessment of the Alliance* (Stanford, CA: Stanford University Press, 1986).

42. Tolman was more complicated in that he posited no reinforcement but something similar ("consequence") that, however, also presupposed a mental state.

43. This was also the argument developed by Skinner. Skinner accepted the concept of reinforcement but avoided framing it in terms of inner states such as drive reduction. To him, a reinforcer was anything that changed the rate with which a response was made; in other words, it was defined retroactively by the behavior it elicited. B. F. Skinner, *Science and Human Behavior* (New York: Macmillan, 1953), 71.

44. Clark L. Hull, *Principles of Behavior: An Introduction to Behavior Theory* (New York: Appleton-Century-Crofts, 1943), 71. On Clark Hull, see Lemov, *World as Laboratory*, chap. 4.

45. Hull, *Principles of Behavior*, chap. 6.

46. Hull, *Principles of Behavior*, 70–71.

47. Whereas Hull specified what he meant by *need*, such as "the need for food, ordinarily called hunger," he specified the term *drive* only through its relationship to *need*: needs "are regarded as producing primary animal *drives*." Hull, *Principles of Behavior*, 57.

48. See Abram Amsel and Jacqueline Roussel, "Motivational Properties of Frustration: I. Effect on a Running Response of the Addition of Frustration to the Motivational Complex," *Journal of Experimental Psychology* 43 (1952): 363–68; Abram Amsel, "The Role of Frustrative Nonreward in Noncontinuous Reward Situations," *Psychological Bulletin* 55, no. 2 (1958): 102–19; Abram Amsel, "Frustrative Nonreward in Partial Reinforcement and Discrimination Learning: Some Recent History and a Theoretical Extension," *Psychological Review* 69, no. 4 (1962): 306–28.

49. Amsel and Roussel, "Motivational Properties of Frustration," 363.

50. For Hull, drives were internal need-related sensory stimuli. He also believed that they could be secondary, that is, produced by external (conditioned) stimuli. Hull, *Principles of Behavior*. See also Sigmund Koch and David E. Leary, *A Century of Psychology as Science* (New York: McGraw-Hill, 1985), 343.

51. Hull, despite certain differences with Tolman, was united with him on the need to introduce a number of what Tolman had called "intervening variables"; these had to be postulated within the Watsonian stimulus-response schema. As for their differences, Hull had an S-R view on learning, whereas Tolman favored a S-S view (latent learning). Further, Hull referred to his behaviorism as "mechanistic," whereas Tolman thought of his as "purposive." Hull and Tolman battled publicly in the 1940s. Hull's theories "won out" over Tolman's, and Hull became very popular in 1940s and 1950s, attracting many students, including O. H. Mowrer and Kenneth

Spence, although B. F. Skinner's ideas ultimately became even more popular. For an account of their differences, see Smith, *Behaviorism and Logical Positivism*.

52. Tolman, "Operational Analysis of 'Demands,'" 383. See also Isaac, *Working Knowledge*, 111.

53. Tolman, "Operational Analysis of 'Demands,'" 388–89.

54. Harry Harlow, "Mice, Monkeys, Men and Motives," *Psychological Review* 60, no. 1 (1953): 23–32. Sometimes a distinction is made between "radical behaviorism" and "methodological behaviorism" to account for these differences, e.g., in B. R. Hergenhahn and Tracy B. Henley, *An Introduction to the History of Psychology*, 7th ed. (Belmont, CA: Cengage Learning, 2013), 395. Not all neobehaviorists were methodological behaviorists, however, as the example of Skinner shows. The use of *classical* versus *neo*behaviorism is useful for periodization.

55. Gordon G. Gallup Jr., "Aggression in Rats as a Function of Frustrative Nonreward in a Straight Alley," *Psychonomic Science* 3 (1965): 99–100.

56. Gordon G. Gallup Jr., "A Technique for Assessing the Motivational Properties of Self-Image Reinforcement in Monkeys" (master's thesis, Washington State University, 1966), 40.

57. Gallup, "Technique for Assessing the Motivational Properties of Self-Image Reinforcement in Monkeys," 1.

58. He never published it but referred to it in his master's thesis: Gallup, "Technique for Assessing the Motivational Properties of Self-Image Reinforcement in Monkeys," 10.

59. Gallup, "Mirror-Image Stimulation and Psychological Research," 15.

60. Gallup, "Aggression in Rats," 99.

61. Gallup, "Mirror-Image Stimulation and Psychological Research," 49; E. H. Hess and J. M. Polt, "Pupil Size as Related to Interest Value of Visual Stimuli," *Science* 132 (1960): 349–50.

62. Gordon G. Gallup Jr., review of *The Unheeded Cry: Animal Consciousness, Animal Pain and Science*, by B. E. Rollin, *Animal Behaviour* 40 (1990): 200–201, on 200.

63. Gallup, "Technique for Assessing the Motivational Properties of Self-Image Reinforcement in Monkeys," 2 (my emphasis).

64. UA MS 198726, box 18, "Summer Appt., NSF Fellow in Psychology, 6–16 to 9-8-66, Scholarship Accts. (17A-7100–5429), $960" Washington State University Archives.

65. This was also published in Gordon G. Gallup Jr., "Mirror-Image Reinforcement in Monkeys," *Psychonomic Science* 5 (1966): 39–40.

66. Gallup, "Mirror-Image Stimulation and Psychological Research," 39.

67. Gallup, "Mirror-Image Stimulation and Psychological Research," 3–4.

68. Gallup, "Mirror-Image Stimulation and Psychological Research," v.

69. Gallup, "Mirror-Image Stimulation and Psychological Research," 2.

70. Gallup, "Mirror-Image Stimulation and Psychological Research," 44.

71. Gallup, "Mirror-Image Stimulation and Psychological Research," 45.

72. Gallup, "Mirror-Image Stimulation and Psychological Research," 44.

73. Gallup, "Mirror-Image Stimulation and Psychological Research," 43.

74. Gallup, "Mirror-Image Stimulation and Psychological Research," 48–50.

75. Other centers were affiliated with institutions including UC Davis; University of Wisconsin, Madison; and Harvard University (at Southborough, MA).

76. Gallup, "Chimpanzees: Self-Recognition," 87.

77. Robert Epstein, Robert P. Lanza, and B. F. Skinner, "'Self-Awareness' in the Pigeon," *Science* 212 (1981): 695–96, on 695.

78. Robert Epstein, "Columban Simulations of Complex Human Behavior" (PhD diss., Harvard University, 1981).

79. *Cognition, Creativity, and Behavior: The Columban Simulations*, featuring and based on work by B. F. Skinner and Robert Epstein, directed and produced by Norman Baxley (Champaign, IL: The Company, 1982), video.

80. He was also adjunct assistant professor at Northeastern University and the University of Massachusetts. Skinner at the time was a professor emeritus of psychology and social relations at Harvard.

81. Epstein did not deny that the self-concept existed in humans, although there as well, he thought it belonged to metaphysics, because one could not scientifically prove its existence.

82. Gordon G. Gallup Jr., "Will Reinforcement Subsume Cognition?," review of *Cognition, Creativity, and Behavior: The Columban Simulations*, PsycCRITIQUES 29, no. 7 (1984): 593–94, on 594.

83. Gordon G. Gallup Jr., "Towards an Operational Definition of Self-Awareness," in *Socioecology and Psychology of Primates*, ed. R. H Tuttle (The Hague: Mouton, 1975), 309–42.

84. In his experiments, Gallup also paid close attention to the socialization of the animal and its effect. This analysis makes sense given the fact that for Gallup the shift in the mirror's stimulus properties would occur only if the animal recognized that it was not a normal social stimulus. For the sake of clarity, I have left that analysis to one side here.

85. Gordon G. Gallup Jr. and Stuart A. Capper, "Preference for Mirror-Image Stimulation in Finches (*Passer domesticus domesticus*) and Parakeets (*Melopsittacus undulates*)," *Animal Behaviour* 18, part 4 (November 1970): 621–24; Gordon G. Gallup Jr. and John Y. Hess, "Preference for Mirror-Image Stimulation in Goldfish (*Carassius auratus*)," *Psychonomic Science* 23 (1971): 63–64.

86. The biologist Nikolaas Tinbergen created stimuli that showed exaggerated features compared to natural stimuli, e.g., artificial eggs with brighter colors. He found that birds preferred the artificial eggs over natural ones. Niko Tinbergen, *The Study of Instinct* (Oxford: Clarendon, 1951).

87. R. Baenninger, "Visual Reinforcement, Habituation, and Prior Social Experience of Siamese Fighting Fish," *Journal of Comparative and Physiological Psychology* 71 (1970): 1–5; J. A. Hogan, "Fighting and Reinforcement in the Siamese Fighting Fish (*Betta splendens*)," *Journal of Comparative and Physiological Psychology* 64 (1967): 356–59; K. B. Melvin and J. E. Anson, "Image-Induced Aggressive Display: Reinforcement in the Paradise Fish," *Psychological Record* 20 (1970): 225–28. The mirror behavior of fighting fish had been known since at least the mid-nineteenth century. In fact, Theodore Cantor's 1849 description of *Betta splendens* included an account of their mirror behavior: *Catalogue of Malayan Fishes* (Calcutta: Baptist Mission, 1849).

88. Gallup and Hess, "Preference for Mirror-Image Stimulation in Goldfish," 64.

89. Gallup and Hess, "Preference for Mirror-Image Stimulation in Goldfish," 63. Gallup also considered, but in this paper rejected, the possibility that, because of the stereotypical nature of fish behavior, it was only through reciprocal exchange with a conspecific that the end of a chain of behavioral events could be reached. For example, a mirror image "never assumes a submissive posture in response to an aggressive display," which may be necessary to "exit from the loop" of behavior and terminate a behavioral episode. But this would have led to habituation, so Gallup rejected the explanation. Gallup and Hess, "Preference for Mirror-Image Stimulation in Goldfish," 64.

90. Gordon G. Gallup Jr., W. A. Montevecchi, and E. T. Swanson, "Motivational Properties of Mirror-Image Stimulation in the Domestic Chicken," *Psychological Record* 22 (1972): 193–99, on 199.

91. Gallup, "Mirror-Image Stimulation and Psychological Research."

92. Gallup, "Chimpanzees: Self-Recognition," 87.

93. Gallup, "Towards an Operational Definition of Self-Awareness," 330.

94. Gallup, "Towards an Operational Definition of Self-Awareness," 329.

95. Such arguments were part of a rising discourse on animal rights; see Edward Baring, "The Human Self: After the Death of Man," in Stefanos Geroulanos, ed., *A Cultural History of Ideas in the Modern Age* (Bloomsbury, forthcoming). However, Gallup did not seem that interested.

96. Gordon G. Gallup Jr. and J. R. Anderson, "Self-Recognition in Animals: Where Do We Stand 50 Years Later? Lessons from Cleaner Wrasse and Other Species," *Psychology of Consciousness: Theory, Research, and Practice* 7, no. 1 (2020): 46–58. In some of these studies, scientists modified the original mirror test to account for self-recognition within a different nonvisual sensory modality, as for instance in the "olfactory mirror" for dogs where dogs investigated their own odor for longer periods of time when it had been modified by an additional odor, thus suggesting that they recognized their own odor. Alexandra Horowitz, "Smelling Themselves: Dogs Investigate Their Own Odours Longer When Modified in an 'Olfactory Mirror' Test," *Behavioural Processes* 143 (2017): 17–24.

97. Peter Singer, *Animal Liberation* (New York: Harper Collins, 1975). In 1981, the philosopher Bernard Rollin noted a growing interest in animal awareness more broadly. Bernard E. Rollin, *Animal Rights and Human Morality* (Buffalo, NY: Prometheus Books, 1981).

98. Diana Reiss, *The Dolphin in the Mirror: Exploring Dolphin Minds and Saving Dolphin Lives* (Boston: Houghton Mifflin Harcourt, 2011), e.g., 23: "Dolphins are among the smartest creatures on the planet. . . . And yet, despite this . . . mankind is slaughtering dolphins at astonishing rates."

99. On the interactions of humans and whales, see D. Graham Burnett, *The Sounding of the Whale: Science and Cetaceans in the Twentieth Century* (Chicago: University of Chicago Press, 2012).

Interlude

1. René Zazzo, *Reflets de miroir et autres doubles* (Paris: Presses universitaires de France, 1993), 13.

2. At the time, Zazzo believed that children generally recognized themselves in the mirror "well before the age of one," but he did not remark on the disparity. Only in 1972, when he returned to his observations, did he note that he had recorded no sign of recognition before that point. Musing that he might have to conclude that his child had developmental difficulties, he wrote, "I was surprised that I hadn't been surprised." Zazzo, *Reflets de miroir*, 14.

3. René Zazzo, "Autobiographie," in *Psychologues de la langue française: Autobiographies*, ed. F. Parot and M. Richelle (Paris: Nathan, 1992), 51–77, on 54.

4. René Zazzo, "La genèse de la conscience de soi (La reconnaissance de soi dans l'image du miroir)," in *Psychologie de la connaissance de soi: Symposium de l'association de psychologie scientifique de langue française (Paris, 1973)* (Paris: Presses universitaires de France, 1975), 145–213, on 146.

5. Zazzo, "La genèse de la conscience de soi," 155–56.

6. Zazzo, "La genèse de la conscience de soi," 152.

7. In 1940, Zazzo was offered a second position, to temporarily replace the psychologist Jean-Maurice Lahy, who fled Paris from the advancing German army, and to protect his position from the seizure of the Vichy regime. When Lahy did not return (he died on his flight from the Germans), Zazzo remained in his dual position, which enabled him to combine clinical work and laboratory research. In 1967, Zazzo was offered another position at the new Nanterre campus of the University of Paris, which he held in parallel until his retirement from all positions in 1980. Zazzo, "Autobiographie."

8. The meetings resulted in the publication of a four volumes entitled *Discussions on Child Development*, one on each meeting. J. M. Tanner and Bärbel Inhelder, eds., *Discussions on Child Development*, 4 vols. (1956; New York: International Universities Press, 1971).

9. René Zazzo, *Les jumeaux, le couple et la personne* (Paris: Presses universitaires de France, 1960), 28.

10. Zazzo, *Les jumeaux*, 28–29.

11. Zazzo, "La genèse de la conscience de soi," 148.

12. Zazzo, "La genèse de la conscience de soi," 149. Here Zazzo again referred to Walter but argued that humans possessed something that Walter's robots did not—"consciousness" (150).

13. Zazzo, "La genèse de la conscience de soi," 156, 158.

14. E.g., Zazzo referred to Amsterdam and Gallup throughout his book *Reflets de miroir*. Zazzo also connected with another disciplinary leader of this time. He worked toward a PhD in 1958 with the Swiss psychologist Jean Piaget (Zazzo was forty-eight years old at the time, and in secure institutional positions). Zazzo presumably took advantage of Piaget's presence in Paris, where he was a professor of genetic psychology at the Sorbonne between 1952 and 1964. Zazzo had interacted with Piaget earlier, at the Study Group on the Psychobiological Development of the Child, where he also met W. Grey Walter. In 1955, Zazzo had traveled with Piaget and Paul Fraisse as a small delegation of leading French psychologists to the USSR, at the invitation of Aleksei Levontiev and Alexander Luria: Jean Piaget, "Some Impressions of a Visit to Soviet Psychologists," *Canadian Psychologist* 5, no. 2 (1956): 32–35. For a comparison of Piaget and Wallon's theories and their impact outside French-speaking psychology, see Beverly Birns, "Piaget and Wallon: Two Giants of Unequal Visibility," in *The World of Henri Wallon*, ed. Gilbert Voyat (New York: Jason Aronson, 1984), 59–69.

15. For Zazzo's and his associates' twin studies, see esp. René Zazzo, "Des jumeaux devant le miroir: Questions de méthode," *Journal de psychologie normale et pathologique* 4 (1975): 389–413; Zazzo, *Reflets de miroir*; Anne-Marie Fontaine, *L'enfant et son image* (Paris: Nathan, 1992); René Zazzo, with Anne-Marie Fontaine, *A travers le miroir—étude sur la découverte de l'image de soi chez l'enfant* (1973, 1 janvier), CERIMES, Canal-U, https://www.canal-u.tv/40869; René Zazzo and Anne-Marie Fontaine, *L'image qui devient un reflect: De l'illusion spéculaire à l'espace des representations*, CNRS audiovisuel (1980).

For earlier twin experiments using an alteration between see-through and mirror situation, see Dorothy Dixon, "The Mirror Behavior of Twins" (MA thesis, University of Florida, 1952); and J. C. Dixon, "Development of Self Recognition," *Journal of Genetic Psychology* 91 (1957): 251–56.

16. Zazzo, "La genèse de la conscience de soi," 151.

17. Zazzo, "La genèse de la conscience de soi," 157–59.

18. Zazzo also used film to present his work to a larger audience, both lay and expert. Of his five films, three were produced by CNRS audiovisuel, the communications unit of the Centre

national de recherche scientifique, whose mission it has been to bring research funded by the CNRS to the public. However, Zazzo's films were also screened and discussed at scientific conferences. For example, Zazzo screened his earliest film, *A travers le miroir: Étude sur la découverte de l'image de soi chez l'enfant* (1973) at the annual meeting of the Association de psychologie scientifique de langue française (which had the theme "Self-Consciousness").

19. Zazzo, "La genèse de la conscience de soi," 180.

20. Zazzo, "La genèse de la conscience de soi," 180.

21. Zazzo, "La genèse de la conscience de soi," 183. Later Zazzo retreated from this hypothesis: it was not through the experience of mirror synchronism that the child learned to recognize herself. Zazzo, *Reflets de miroir*.

22. Zazzo, "La genèse de la conscience de soi," 188.

23. Zazzo, "La genèse de la conscience de soi," 176.

24. Zazzo, "La genèse de la conscience de soi," 175.

25. Zazzo, "La genèse de la conscience de soi," 148.

26. Zazzo, "La genèse de la conscience de soi," 179.

27. Zazzo, "Autobiographie," 53.

28. See, e.g., Zazzo, "Autobiographie," 52.

29. Zazzo, "La genèse de la conscience de soi," 175. Lacan had placed mirror recognition at six months. And he took this from an already challenged dating in Preyer and Wallon derived ultimately from Darwin. In the conversation about the paper, the psychoanalyst Odette Brunet pushed back. A few days before, she attested, she had held a baby of four months to the mirror, and "I have witnessed exactly the scene of jubilation of which Lacan speaks." The baby, she said, "has laughed as never before at anybody, [and] I had the impression that he know[s] it was him." Quoted in Zazzo, "La genèse de la conscience de soi," 209.

Chapter 5: The Mirror Test That Never Happened

1. As we will see, the mirror stage did not require for Lacan an actual mirror. An ego could also be formed when a child identified with another child.

2. Jacques Lacan, "The Mirror Stage as Formative of the I Function as Revealed in Psychoanalytic Experience" (1949), in *Écrits: The First Complete Edition in English*, trans. Bruce Fink (New York: W. W. Norton, 2006), 75–81, on 76.

3. There is not much recent historical work on Lacan, but my chapter builds on works that emphasize the clinical and practical aspects of Lacan's work, such as Dany Nobus, *Jacques Lacan and the Freudian Practice of Psychoanalysis* (London: Routledge, 2000); and Bruce Fink, *A Clinical Introduction to Lacanian Psychoanalysis: Theory and Technique* (Cambridge, MA: Harvard University Press, 1997).

4. Jacques Lacan, "Some Reflections on the Ego," *International Journal of Psycho-analysis* 34 (1953): 11–17, on 14. The paper was read by Lacan at the British Psycho-analytic Society Congress on May 2, 1951.

5. René Zazzo, *Reflets de miroir et autres doubles* (Paris: Presses universitaires de France, 1993), 175.

6. Bertrand Ogilvie discounts the debates over dating, by suggesting that it is the behavior of the child that matters, rather than its result: *Lacan: La formation du concept de sujet, 1932–1949*

(Paris: Presses universitaires de France, 1987), 103–4. But he ignores the fact that the dating determines the relative order of behaviors, which is essential for Lacan's interpretation.

7. Recall the air of generality surrounding the scene Lacan describes—the child could be held up by an adult or a walker—which detaches the event from any particular instance.

8. Henri Wallon, *Les origines du caractère chez l'enfant: Les préludes du sentiment de personnalité* (1934; Paris: Presses universitaires de France, 2009), 184.

9. Wallon, *Les origines*, 7.

10. Wallon, *Les origines*, 8.

11. Wallon, *Les origines*, 8.

12. Wallon, *Les origines*, 23. He referred here to his 1925 work *L'enfant turbulent: Etude sur les retards et les anomalies du développement moteur et mental* (Paris: Alcan, 1925).

13. Wallon, *Les origines*, 208–9.

14. Wallon, *Les origines*, 195.

15. Wallon, *Les origines*, 198–99; see also 342.

16. Wallon, *Les origines*, 206. Here Wallon made a brief mention of the mirror, though he was keen to distinguish this from the real business of the mirror. In recognizing the similarities between what chimpanzees and one-year-old children did with the mirror, Wallon was clear that the "essential motive for their behavior was the mirror and not their image" (233). On the "chimpanzee stage," see Karl Bühler, *Die geistige Entwicklung des Kindes*, 3rd ed. (1918; Jena: Fischer, 1922), 82.

17. See Shuli Barzilai, "Models of Reflexive Recognition. Wallon's *Origines du caractère* and Lacan's 'Mirror Stage,'" *Psychoanalytic Study of the Child* 50 (1995): 368–82. I differ, however, in my reading of how Wallon treated the differences between humans and animals.

18. Wallon, *Les origines*, 218–19.

19. Note the conflation of self and other here.

20. Wallon, *Les origines*, 220.

21. Wallon, *Les origines*, 221.

22. Wallon, *Les origines*, 221.

23. Wallon, *Les origines*, 223.

24. Wallon, *Les origines*, 224.

25. Wallon, *Les origines*, 227.

26. He later revised the dating to two years: Henri Wallon, "Kinesthésie et image visuelle du corps propre chez l'enfant," *Bulletin de psychologie* 7, no. 5 (1954): 239–46.

27. Wallon, *Les origines*, 225.

28. Wallon, *Les origines*, 226.

29. Wallon, *Les origines*, 226.

30. Wallon, *Les origines*, 230.

31. Wallon, *Les origines*, 227.

32. Wallon, *Les origines*, 230.

33. Guillaume's book, a baby diary of sorts, was one of the sources that Wallon considered and built his theories on. Paul Guillaume, *L'imitation chez l'enfant: Étude psychologique* (Paris: Alcan, 1925).

34. Wallon, *Les origines*, 231.

35. Wallon, *Les origines*, 232.

36. Wallon, *Les origines*, 234.

37. Wallon, *Les origines*, 235.

38. Wallon, *Les origines*, 254.

39. Wallon, *Les origines*, 257.

40. Wallon, *Les origines*, 231.

41. Wallon, *Les origines*, 231.

42. Henri Wallon, "The Psychological Development of the Child," in Gilbert Voyat, *The World of Henri Wallon* (New York: Jason Aronson, 1984), 133–46, on 140. It is reprinted in *International Journal of Mental Health* 1, no. 4 (1973): 29–39; it was based on extracts from *L'évolution psychologique de l'enfant* (Paris: Armand Calin, 1941).

43. Henri Wallon, "Genetic psychology," in Voyat, *World of Henri Wallon*, 15–32, on 17–18. It was first published as "La psychologie génétique," *Bulletin de psychologie* 10, no. 1 (1956): 3–10; then reprinted in *Enfance* 12, nos. 3–4 (1973): 220–34.

44. Henri Wallon, "Psychological Development," 138–39.

45. Jacques Lacan, "Le stade du miroir comme formateur de la fonction du je, telle qu'elle nous est révélée dans l'expérience psychanalytique," *Revue française de psychanalyse* 13, no. 4 (1949): 449–55. On the history of the paper, see Elisabeth Roudinesco, "The Mirror Stage: An Obliterated Archive," in *The Cambridge Companion to Lacan*, ed. Jean-Michel Rabaté (Cambridge: Cambridge University Press, 2003), 25–34.

46. Jacques Lacan, *De la psychose paranoïaque dans les rapports avec la personnalité* (Paris: E. Le François, 1932; Paris: Le Seuil, 1975).

47. After the attack, Anzieu was taken to the police station, from there to the special infirmary and the women's prison at Saint-Lazare, where she fell into a delusional state. After three weeks, she was sent to the Sainte-Anne hospital, where she became Lacan's patient. Elisabeth Roudinesco, *Jacques Lacan & Co.: A History of Psychoanalysis in France, 1925–1985* (Chicago: University of Chicago Press, 1990), 33.

48. Quoted in Roudinesco, *Jacques Lacan & Co.*, 114.

49. Note that his training analysis was a huge failure and came to a halt in 1953. Lacan was made a full member of the SPP under the promise to finish his analysis with Loewenstein—but he didn't keep his promise. Roudinesco, *Jacques Lacan & Co.*, 122.

50. Lacan recalled the incident in "Presentation on Psychical Causality," in *Écrits*, 123–58, on 150–51.

51. Roudinesco, "Mirror Stage," 25.

52. It was listed as "Dr. J. Lacan (Paris). The Looking-Glass Phase," in "Report of the Fourteenth International Psycho-analytical Congress," *International Journal of Psycho-analysis* 18 (1937): 72–107, on 78.

53. Roudinesco, *Jacques Lacan & Co.*, 142.

54. On his reception and reaction at the congress, see Roudinesco, "Mirror Stage," 25–27. See also Roudinesco, *Jacques Lacan & Co.*, esp. chap. 1.

55. Roudinesco, "Mirror Stage," 27. There are a number of accounts of the differences between Wallon and Lacan, including Barzilai, "Models of Reflexive Recognition." The most extensive analyses can be found in Emile Jalley, *Freud, Wallon, Lacan: L'enfant au miroir* (Paris: EPEL, 1998); and Juan Pablo Lucchelli, *Lacan: De Wallon à Kojève* (Paris: Edition Michèle, 2017). Michael Billig gives a "rhetorical analysis" of Lacan's misreadings and miscitations of other

psychological figures, noting also the paucity of citations of Wallon: "Lacan's Misuse of Psychology: Evidence, Rhetoric and the Mirror Stage," *Theory, Culture and Society* 23, no. 4 (2006): 1–26.

56. Jacques Lacan, "Les complexes familiaux dans la formation de l'individu: Essay d'analyse d'une function en psychologie," in *L'encyclopédie française*, vol. 8, *La vie mentale*, ed. Henri Wallon (Paris: Société nouvelle de l'encyclopédie française, 1938), 23–78, on 33, 40, 45. The text was later republished in Lacan, *Autres écrits* (Paris: Le Seuil, 2001), 23–84. See also the discussion of the Oedipus complex in Lacan, "Mirror Stage," 99.

57. Lacan, "Les complexes familiaux," 33, 38.

58. Lacan, "Les complexes familiaux," 37.

59. Lacan, "Les complexes familiaux," 37–38.

60. Lacan, "Les complexes familiaux," 42.

61. Lacan, "Les complexes familiaux," 43–44.

62. Lacan, "Les complexes familiaux," 41. See also Wallon, *Les origines*, 233.

63. Lacan, "Les complexes familiaux," 43.

64. Lacan, "Les complexes familiaux," 43.

65. Lacan, "Some Reflections on the Ego," 14. Lacan said here that the interest is shown "over eight months," but it is likely that (writing in this case in English) he made an error and meant *at* eight months.

66. Lacan attributed the dating of six months to Baldwin, but scholars have been unable to locate the reference. René Zazzo, "La genèse de la conscience de soi (La reconnaissance de soi dans l'image du miroir)," in *Psychologie de la connaissance de soi: Symposium de l'association de psychologie scientifique le langue française (Paris, 1973)* (Paris: Presses universitaires de France, 1975), 145–213, on 175. Many think that he might be referring to Darwin, who dated the emotional response at six months. Billig, "Lacan's Misuse of Psychology," 10–11; Zazzo, "La genèse de la conscience de soi," 175.

67. Lacan, "Les complexes familiaux," 41. I would like to thank Yael Geller for helping me understand the differences between specular identification and consciousness of the self.

68. He did suggest that there was vast variability in development due to differences in family structure, and therefore the presence or not of viable rivals. Lacan, "Les complexes familiaux," 37.

69. Lacan, "Les complexes familiaux," 42.

70. Wallon, *Les origines*, 213.

71. Lacan, "Les complexes familiaux," 42 (my emphasis).

72. Wallon, *Les origines*, 199.

73. Lacan, "Les complexes familiaux," 32.

74. Lacan praised the existentialists for recognizing the role of negativity but criticized them for yoking that to the unity and autonomy of the ego rather than being that which undermined it. Lacan, "Mirror Stage," 79–80.

75. Lacan, "Mirror Stage," 78.

76. Lacan, "Mirror Stage," 75–76 (bracketed interpolation in original).

77. Nevertheless, Lacan argued that this stage "is characterized by transformations of nervous structure quick and profound enough to control [*dominer*] individual differentiations." Lacan, "Les complexes familiaux," 38.

78. Wallon, *Les origines*, 230.

79. Lacan, "Les complexes familiaux," 42–43.

80. See Billig, "Lacan's Misuse of Psychology," 14.

81. Lacan, "Les complexes familiaux," 41–42.

82. Mikkel Borch-Jacobsen sets up a similar duality in his book, arguing that given Lacan's account of sociability, there were residues of Wallon's other version in his 1938 article. Mikkel Borch-Jacobsen, *Lacan: The Absolute Master* (Stanford, CA: Stanford University Press, 1991), 67–71.

83. Lacan, "Les complexes familiaux," 34.

84. Lacan, "Les complexes familiaux," 40; Lacan, "Mirror Stage," 79.

85. Lacan, "Presentation on Psychical Causality," 151.

86. Lacan, "Les complexes familiaux," 42.

87. Sigmund Freud, "The 'Uncanny' (1919)," in *The Standard Edition of the Complete Psychological Works of Sigmund Freud*, ed. James Strachey, vol. 17, *"An Infantile Neurosis, and Other Works" (1917–1919)* (London: Hogarth, 1955), 217–56. It was first published as "Das Unheimliche," *Imago* 5, nos. 5–6 (1919): 297–324. On the double and psychoanalysis, see Otto Rank, *The Double: A Psychoanalytic Study* (1925; London: Karnac, 1989).

88. Rank quoted in Freud, "'Uncanny,'" 235.

89. Freud, "'Uncanny,'" 235.

90. Freud, "'Uncanny,'" 235. When Freud caught sight of his reflected image on a train, he "thoroughly disliked his appearance" (247n1).

91. Freud, "'Uncanny,'" 235.

92. Sigmund Freud, "On Narcissism: An Introduction (1914)," in *The Standard Edition*, vol. 14, *"On the History of the Psycho-Analytic Movement," "Papers on Metapsychology," and Other Works (1914–1916)* (London: Hogarth, 1957), 67–102, on 94. It was first published as "Zur Einführung des Narzißmus," *Jahrbuch der Psychoanalyse* 6 (1914): 1–24.

93. Borch-Jacobsen makes the argument that Lacan's early reading of Freud and especially Freud's theory of narcissism actually betrayed Lacan's Hegelian leanings. Borch-Jacobsen, *Lacan*, 28–29, 50.

94. Lacan, "Mirror Stage," 79.

95. Jean Laplanche and Jean-Bertrand Pontalis make a similar shift in their article on narcissism, though they argue that it was implicit if never fully developed in Freud. J. Laplanche and J.-B. Pontalis, *The Language of Psycho-analysis*, trans. Donald Nicholson-Smith (London: Hogarth, 1973), s.v. "Narcissism," 255–57.

96. See, e.g., Jacques Lacan, *The Seminar of Jacques Lacan*, ed. Jacques-Alain Miller, bk. 1, *Freud's Papers on Technique, 1953–1954*, trans. John Forrester (1975; New York: W. W. Norton, 1988), 129–42.

97. It also followed in part from Lacan's participation in Alexandre Kojève's famous seminars on Hegel at the École pratique des hautes études, which emphasized the question of desire from nothing in a discussion of self-consciousness. See Roudinesco, *Jacques Lacan & Co.*, 141.

98. Lacan's so-called Rome Report was a paper delivered at a conference at the Institute of Psychology at the University of Rome on September 26–27, 1953, and published as Jacques Lacan, "The Function and Field of Speech and Language in Psychoanalysis," in *Écrits*, 197–268.

99. Lacan, "Function and Field of Speech," 223.

100. On the talking cure and a discussion of the place of "having a conversation" in turn-of-the-century Viennese medicine and broader culture, see Joshua Bauchner, "The Lives of the Mind: Scientific Concept and Everyday Experience from Psychophysics to Psychoanalysis" (PhD diss., Princeton University, 2021).

101. Lacan, "Function and Field of Speech," 216.

102. Lacan, "Function and Field of Speech," 219.

103. Lacan, *Seminar*, bk. 1, 138. In the Rome Report, Lacan clarified that animals used symbols too but they were not yet structured like a language, since their relationship to the signified was fixed. Lacan, "Function and Field of Speech," 228.

104. That is why, as Elisabeth Roudinesco has insisted, the mirror stage was ultimately for Lacan not a passing moment in human development. Roudinesco, *Jacques Lacan & Co.*, 142–47.

105. Jacques Lacan, "Seminar on 'The Purloined Letter,'" in *Écrits*, 6–48, on 40.

106. Lacan, "Function and Field of Speech," 250.

107. Jacques-Alain Miller, "A Critical Reading of Jacques Lacan's *Les complexes familiaux*," trans. Thomas Svolos, *Lacanian International Review*, summer 2005, https://www.lacan.com/frameX14.htm; Roudinesco, *Jacques Lacan & Co.*; Jalley, *Freud, Wallon, Lacan*.

108. Miller, "Critical Reading."

109. Miller, "Critical Reading."

110. Shuli Barzilai notes the difference in Wallon's and Lacan's accounts on this point but misses the fact that the reason Wallon felt compelled to appeal to the symbolic was present for Lacan too. Barzilai, "Models of Reflexive Recognition,"

111. Sigmund Freud, "Beyond the Pleasure Principle (1920)," in *The Standard Edition*, vol. 18, *"Beyond the Pleasure Principle," "Group Psychology," and Other Works (1920–1922)* (London: Hogarth, 1955), 1–64, on 15.

112. Lacan, "Seminar on 'The Purloined Letter,'" 35.

113. Lacan, "Function and Field of Speech," 228 (bracketed interpolation in original).

114. Lacan, "Function and Field of Speech," 229.

115. Henri Bouasse, *Optique géométrique élémentaire: Focométrie, optométrie* (Paris: Delagrave, 1917).

116. Lacan, *Seminar*, bk. 1, 78–79. Later Lacan flipped the experiment so it was the vase that was inverted in the box; this highlighted the imaginary nature of the ego and the reality of desires (123–24).

117. Lacan *Seminar*, bk. 1, 74.

118. Lacan, *Seminar*, bk. 1, 80.

119. See also Lacan, *Seminar*, bk. 1, 140–41.

120. Jacques Lacan, *The Seminar of Jacques Lacan*, ed. Jacques-Alain Miller, bk. 10, *Anxiety*, trans. A. R. Price (Cambridge, UK: Polity, 2014), 30.

121. Lacan, *Seminar*, bk. 10, 32 (italics in original).

122. Lacan, *Seminar*, bk. 10, 32.

123. Lacan, "Mirror Stage," 76.

124. Lacan, "Presentation on Psychical Causality," 149.

125. Lacan, "Presentation on Psychical Causality," 150.

126. Lacan, *Seminar*, bk. 1, 119.

127. Lacan, "Some Reflections on the Ego," 11.

128. Lacan, "Mirror Stage," 79 (emphasis in original).

129. For the temporal complications that followed from this, see Jane Gallop, "Lacan's 'Mirror Stage': Where to Begin," *SubStance* 11/12, no. 4/1 (1982–83): 118–28.

130. Lacan, "Function and Field of Speech," 213.

131. Lacan, "Function and Field of Speech," 211–13.

132. Lacan, "Presentation on Psychical Causality," 149–50.

Chapter 6: There Are No Mirrors in New Guinea

1. Letter, Edmund Carpenter to Marshall McLuhan, October 15, 1969, Papers of Edmund Snow Carpenter, Rock Foundation, New York. The Carpenter Papers are now at the National Anthropology Archives at the Smithsonian Institution.

2. Edmund Carpenter, "The Tribal Terror of Self-Awareness," in *Principles of Visual Anthropology*, 2nd rev. ed., ed. Paul Hockings (1975; Berlin: Mouton de Gruyter, 1995), 481–91, on 482.

3. The thought experiment is similar to Avicenna's "floating man" who did not require sense perception to build a sense of self, even though perception was the normal route.

4. René Zazzo, "La genèse de la conscience de soi (La reconnaissance de soi dans l'image du miroir)," in *Psychologie de la connaissance de soi: Symposium de l'association de psychologie scientifique de langue française (Paris, 1973)* (Paris: Presses universitaires de France, 1975), 145–213, on 161–62.

5. Edmund Carpenter, *Oh, What a Blow That Phantom Gave Me!* (New York: Holt, Rinehart and Winston, 1972), 113. Although apparently not part of Carpenter's official mission, there was a problem with cannibalism among the Biami that the Australian government sought to control. Carpenter made recommendations similar to those discussed later in the chapter (writing the accused natives' names along with Polaroid shots into census book, and showing them the results). Letter, Carpenter to parent-in-law, December 31, 1969, Papers of Edmund Snow Carpenter.

6. Carpenter, "Tribal Terror," 482.

7. His wide range of publications on the topic is listed in Harald E. L. Prins and John Bishop, "Edmund Carpenter: Explorations in Media and Anthropology," *Visual Anthropology Review* 17, no. 2 (2001–2): 110–40.

8. One of the outcomes was the award-winning short film *College* (1964; nineteen minutes, 16 mm) produced in collaboration with Jacob Bronowski.

9. According to Carpenter, the school never existed: "It was simply a bunch of islanders watching the greatest show on earth. A table in the museum coffee shop served as meeting place." Edmund Carpenter, "That Not-So-Silent Sea," appendix to *The Virtual Marshall McLuhan*, by Donald F. Theall (Montreal: McGill-Queen's University Press, 2001), 236–61, on 251. But there was a coherent set of ideas that all members shared; their thinking was also informed by the work of Dorothy Lee. On the effect of language on identity and sociality, see Dorothy Lee, "Being and Value in Primitive Culture," *Journal of Philosophy* 46 (1949): 401–15; see also Benjamin Lee Whorf, *Language, Thought and Reality* (Cambridge, MA: MIT Press, 1956). On the Toronto group and *Explorations*, see Michael Darroch "Bridging Urban and Media Studies: Jaqueline Tyrwhitt and the Explorations Group, 1951–1957," *Canadian Journal of Communication* 33, no. 2 (2008): 147–69; Michael Darroch, "Sigfried Giedion und die *Explorations*: Die

anonyme Geschichte der Medien-Architektur," *Zeitschrift für Medienwissenschaft* 11, no. 2 (2014): 144–54.

10. Carpenter, "Tribal Terror," 481.

11. Marshall McLuhan, *The Gutenberg Galaxy: The Making of Typographic Man* (1962; Toronto: University of Toronto Press, 2017), e.g., 26.

12. McLuhan, *Gutenberg Galaxy*, 286.

13. McLuhan, *Gutenberg Galaxy*, 286.

14. Edmund Carpenter and Ken Heyman, *They Became What They Beheld* (New York: Outerbridge and Dienstfrey / Ballantine Books, 1970). Note that the book—in correspondence with its aphoristic, nonlinear style—does not have page numbers.

15. McLuhan, *Gutenberg Galaxy*, 248. For other accounts on how media have shaped the self-other relationship, see esp. Lisa Gitelman, *Scripts, Grooves and Writing Machines: Representing Technology in the Edison Era* (Stanford, CA: Stanford University Press, 1999); Lisa Gitelman, *Always Already New: Media, History and the Data of Culture* (Cambridge, MA: MIT Press, 2006); Friedrich Kittler, *Gramophone, Film, Typewriter* (Stanford, CA: Stanford University Press, 1999); Jeffrey Sconce, *Haunted Media: Electronic Media from Telegraphy to Television* (Durham, NC: Duke University Press, 2000).

16. The two men's shared optimism about new media was also expressed in their teaching concerns. See, for example, their 1957 "Classrooms without Walls," reprinted in *Explorations in Communication: An Anthology*, ed. Edmund Carpenter and Marshall McLuhan (Boston: Beacon, 1960), 1–3.

17. Marshall McLuhan, *The Mechanical Bride: Folklore of Industrial Man* (Berkeley, CA: Gingko, 1951), v.

18. In discussing electronic media, McLuhan drew on the myth of Narcissus, and so on reflections, but importantly for him, the mirror here was not a tool for self-recognition. Marshall McLuhan, *Understanding Media: The Extensions of Man* (1964; Cambridge, MA: MIT Press, 1994), chap. 4.

19. "A Candid Conversation with Marshall McLuhan, the High Priest of Popcult and Metaphysician of Media," in *Voices of Concern: The Playboy College Reader*, ed. the Editors of *Playboy* (New York: Harcourt Brace Jovanovich, 1971): 448–88. Reprinted from *Playboy* magazine, March 1969.

20. "Candid Conversation," 462. On the meaning of *tactile*, see Philip Marchand, *Marshall McLuhan: The Medium and the Messenger, a Biography* (1989; Cambridge, MA: MIT Press, 1998), 153. On the history and philosophy of tactile media, see Henning Schmidgen, *Horn oder Die Gegenseite der Medien* (Berlin: Matthes und Seitz, 2018), esp. chap. 4.

21. McLuhan, *Understanding Media*, 22. The effect of a cold medium was context dependent. It was beneficial in a hot, industrial, society because of its cooling effects. Nevertheless, McLuhan argued that a cold medium could be destructive in a cold, tribal society, for accentuating preexisting tendencies.

22. "Candid Conversation," 462.

23. "Candid Conversation," 477.

24. "Candid Conversation," 477.

25. Carpenter and Heyman, *They Became What They Beheld*.

26. Carpenter and Heyman, *They Became What They Beheld*.

27. Carpenter and Heyman, *They Became What They Beheld*.

28. Edmund Carpenter, "Space Concepts of the Aivilik Eskimos," *Explorations* 5 (1955): 131–45. Elsewhere, Carpenter notes the overlap with Western art that tried to recuperate such a sense of space, e.g., in the work of Paul Klee and Joan Miró.

29. Edmund Carpenter, interview, in *Oh, What a Blow That Phantom Gave Me!*, dir. John M. Bishop and Harald E. L. Prins (Watertown, MA: Documentary Educational Resources, 2003), video, at 8:08.

30. Carpenter, "Space Concepts," 144.

31. Edmund Carpenter, *Eskimo Realities* (New York: Holt, Rinehart and Winston, 1973), 195. Carpenter used the colonial name Eskimo to refer to both the Iglulingmiut and the Aivilingmiut.

32. In his initially positive assessment of film, Carpenter differed from McLuhan, for whom film was a hot medium. Note, however, that for McLuhan not all films were created equal: silent movies were better than talkies because they required the viewers to fill in the sound, thus making them more active.

33. Carpenter mentioned the film several times; see, e.g., Edmund Carpenter, "Television Meets the Stone Age," *TV Guide* 19, no. 3 (1971): 14–16.

34. Éclair advert, Papers of Edmund Snow Carpenter.

35. Edmund Carpenter, "Frank Speck, Quiet Listener," in *The Life and Times of Frank G. Speck, 1881–1950*, ed. Roy Blankenship (Philadelphia: University of Pennsylvania Publications in Anthropology, 1991), 78–83.

36. Edmund Carpenter, "Arctic Witnesses," in *Fifty Years of Arctic Research: Anthropological Studies from Greenland to Siberia*, ed. Rolf Gilberg and Hans Christian Gullov, Publications of the National Museum Ethnographical Series 18 (Copenhagen: National Museum of Denmark, Department of Ethnography, 1997), 303–10. On *Nanook of the North* and the filmic representation of non-Western indigenous people, see Fatimah Tobing Rony, *The Third Eye: Race, Cinema, and Ethnographic Spectacle* (Durham, NC: Duke University Press, 1996), esp. chap. 4; see also Karl Heider, *Seeing Anthropology: Cultural Anthropology through Film* (Boston: Allyn and Bacon, 2001). For a history of visual anthropology, see Anna Grimshaw, "Visual Anthropology," in *A New History of Anthropology*, ed. Henrika Kucklick (Malden, MA: Blackwell, 2008), 293–309.

37. Edmund Carpenter, interview by Bunny McBride, in McBride, "A Sense of Proportion: Balancing Subjectivity and Objectivity in Anthropology" (MA thesis, Columbia University, 1980), 110.

38. Edmund Carpenter, Frederick Varley, and Robert Flaherty, *Eskimo* (Toronto: University of Toronto Press, 1959); the book also served as *Explorations* 9.

39. Carpenter, *Oh, What a Blow That Phantom Gave Me!*, 180.

40. Letter, Carpenter to McLuhan, September 23, 1969, Papers of Edmund Snow Carpenter.

41. Carpenter, CBC Interview, 5, Papers of Edmund Snow Carpenter.

42. The quote in this section's heading is from Carpenter in *Oh, What a Blow That Phantom Gave Me!*, 115. This quote was also scribbled above the patrol reports, Papers of Edmund Snow Carpenter.

43. Carpenter, "Tribal Terror," 488.

44. The core of the book was written before Carpenter's trip to Papua New Guinea, based on an earlier *Harper's Bazaar* piece under McLuhan's name, although parts were added in after the trip. However, we do not yet find a negative assessment of electronic media. Marshall McLuhan, "Fashion Is Language," *Harper's Bazaar* 101, no. 3077 (April 1968): 150–67.

45. Carpenter and Heyman, *They Became What They Beheld.*

46. Similarly, McLuhan, although identifying both the birth of phonetic writing and the development of the printing press with specific historical moments, took inspiration from the Cadmus myth, first invoked by another Toronto school member, Harold Innis: "The Greek King Cadmus, who introduced the phonetic alphabet to Greece, was said to have sown the dragon's teeth and that they sprang up armed men." McLuhan, *Gutenberg Galaxy*, 56.

47. Carpenter, "Television Meets the Stone Age."

48. Carpenter, "Tribal Terror," 481.

49. Carpenter, "Tribal Terror," 481. The languages in Papua New Guinea form two main groups: The Austronesian languages (including Enga, Kuanua, Melpa, Kuman, and Huli) are spoken by about a sixth of the population. The rest of the population speak the other seven hundred or so languages. In addition, two pidgin languages, Hiri Motu and Tok Pisin, developed with the advent of colonialism. Tok Pisin has become the national lingua franca. Angela M Gilliam, "Language and 'Development' in Papua New Guinea," *Dialectical Anthropology* 8, no. 4 (1984): 303–18.

50. Carpenter was wrong about the entirely isolated nature of the Papuan tribes. Papua New Guinea, including the more-difficult-to-access highlands, had been the target of many outsiders since the late nineteenth century, including gold prospectors, patrol officers, missionaries, and anthropologists. The last came in three waves, in the 1930s, the 1950s, and the 1970s (Carpenter's wave). See Terence E. Hays, ed., *Ethnographic Presents: Pioneering Anthropologists in the Papua New Guinea Highlands*, Studies in Melanesian Anthropology 12 (Berkeley: University of California Press, 1992), esp. chap. 1.

51. They have also been reproduced in Carpenter, *Oh, What a Blow That Phantom Gave Me!*

52. Government Patrol Reports, 1962, 1960, Papers of Edmund Snow Carpenter.

53. Carpenter, "Tribal Terror," 488.

54. Carpenter, "Tribal Terror," 489.

55. Letter, Carpenter to DeMenil family, December 31, 1969, Papers of Edmund Snow Carpenter.

56. Carpenter, *Oh, What a Blow That Phantom Gave Me!*, 48.

57. See, e.g., Charles W. Abel, *Savage Life in New Guinea: The Papuan in Many Moods* (London: London Missionary Society, [1902?]), 32; Sidney Spencer Broomfield, *Karachola; or, The Mighty Hunter: The Early Life and Adventures of Sidney Spencer Broomfield* (New York: W. Morrow, 1931), 164; Antwerp Edgar Pratt, *Two Years among New Guinea Cannibals: A Naturalist's Sojourn among the Aborigines of Unexplored New Guinea* (Philadelphia: Lippincott, 1906). See also Edward L. Schieffelin and Robert Crittenden, *Like People You See in a Dream: First Contact in Six Papuan Societies* (Stanford, CA: Stanford University Press, 1991).

58. Jack Hides, *Papuan Wonderland* (Glasgow: Blackie and Son, 1936), 85. Note that Puya Indane also rejected other trade objects Hides presented to him, such as beads or knives, much to the Hides's annoyance.

59. Carpenter, *Oh, What a Blow That Phantom Gave Me!*, 120.

60. McLuhan, to my knowledge, never classified mirrors along the hot-cold axis but had the following to say about them: "The youth Narcissus mistook his own reflection in the water for another person. This extension of himself by mirror numbed his perceptions until he became the servomechanism of his own extended or repeated image." McLuhan, *Understanding Media*, 41.

61. Carpenter, "Tribal Terror," 483. Note that this, too, corresponded to a trope in the early anthropological literature, to note the shock at the sight of their selves in the mirror. Ilana Gershon, "Mirrors and Numbers among Others: Technologies of Identification in Papua New Guinea," *Paideuma* 54 (2008): 85–108.

62. Carpenter, "Tribal Terror," 485.

63. Carpenter, CBC Interview.

64. Carpenter, *Oh, What a Blow That Phantom Gave Me!*, 124. "Vomited up an organ" was a further example. Carpenter, "Tribal Terror," 483.

65. Carpenter, "Tribal Terror," 485.

66. Carpenter, *Oh, What a Blow That Phantom Gave Me!*, 129.

67. Carpenter, *Oh, What a Blow That Phantom Gave Me!*, 130.

68. This is evident from the film that Carpenter's team took of this encounter, included in Bishop and Prins, *Oh, What a Blow That Phantom Gave Me!* It is of course unclear whether it was the sound of his voice that caused embarrassment or the larger situation to which Carpenter and his crew subjected the man.

69. Carpenter, CBC Interview.

70. Carpenter, *Oh, What a Blow That Phantom Gave Me!*, 130. On technologies of identification in Papua New Guinea, see Gershon, "Mirrors and Numbers." On technologies of identification in other contexts, see esp. Jean Comaroff and John Comaroff, *Of Revelation and Revolution*, vol. 1, *Christianity, Colonialism and Consciousness in South Africa* (Chicago: University of Chicago Press, 1991); Andrew Lattas, "Technologies of Visibility: The Utopian Politics of Cameras, Televisions, Videos and Dreams in New Britain," *Australian Journal of Anthropology* 17, no. 1 (2006): 15–31. For excellent overviews of the literature on colonialism, decolonization, and technology, see David Arnold, "Europe, Technology, and Colonialism in the 20th Century," *History and Technology* 21, no. 1 (2005): 85–106; Warwick Anderson, introduction to "Postcolonial Technoscience," special issue, *Social Studies of Science* 32, no. 5/6 (2002): 643–58.

71. Carpenter, "Television Meets the Stone Age," 16.

72. Carpenter, *Oh, What a Blow That Phantom Gave Me!*, 130.

73. Carpenter, *Oh, What a Blow That Phantom Gave Me!*, 130.

74. Carpenter, "Television Meets the Stone Age," 15.

75. Carpenter, interview, in Bishop and Prins, *Oh, What a Blow That Phantom Gave Me!*

76. Carpenter, "Television Meets the Stone Age," 16.

77. Carpenter, "Television Meets the Stone Age," 16.

78. Carpenter's emphasis. Letter, Carpenter to McLuhan, September 23, 1969, Papers of Edmund Snow Carpenter. This was true both when they were being filmed, and when they were filming themselves—the two for Carpenter had the same effect.

79. Carpenter, *Oh, What a Blow That Phantom Gave Me!*, 119.

80. Edward Carpenter interview with Bunny McBride, 1980. Many thanks to Harald Prins and Bunny McBride for sharing the interview transcript with me.

81. Carpenter, *Oh, What a Blow That Phantom Gave Me!*, 119.

82. See note 70.

83. Prins and Bishop, "Edmund Carpenter," 134.

84. Carpenter, interview, in Bishop and Prins, *Oh, What a Blow That Phantom Gave Me!*

85. Carpenter, *Oh, What a Blow That Phantom Gave Me!*, 139.

86. Carpenter, *Oh, What a Blow That Phantom Gave Me!*, 99.

87. Prins and Bishop, "Edmund Carpenter," 131.

88. Or with some parts of it. It is quite possible that Carpenter, as someone who had emphasized the perceptual dimensions of anthropology (his early studies on spatial orientation, his media experiments), was discouraged by the "textual revolution" of proponents such as Clifford Geertz, even if he might have shared their self-criticism. On the oppositions between perceptual and textual anthropology, see David Howes, "Controlling Textuality: A Call for a Return to the Senses," *Anthropologica* 32, no. 1 (1990): 55–73.

89. Note that Carpenter chose McLuhan's "tribal man" over Rousseau's "noble savage," which he used to criticize the Western "search for the primitive . . . to play the role of alter ego for us." Carpenter, *Oh, What a Blow That Phantom Gave Me!*, 102. This critique should of course include himself (and sometimes did).

90. Carl Schuster and Edmund Snow Carpenter, *Materials for the Study of Social Symbolism in Ancient and Tribal Art: A Record of Tradition and Continuity*, ed. Edmund Snow Carpenter and Lorraine Spiess, 3 vols., 12 bks. (New York: Rock Foundation, 1986–88).

91. Schuster and Carpenter, *Materials*, vol. 1, 40–41.

92. Schuster and Carpenter, *Materials*, vol. 1, 11.

93. Schuster and Carpenter, *Materials*, vol. 3, part 1, 30, 23, 27

94. Schuster and Carpenter, *Materials*, vol. 1, 821.

95. Carpenter, *Oh, What a Blow That Phantom Gave Me!*, 101–2.

96. Adrianna Link, "Documenting Human Nature: E. Richard Sorenson and the National Anthropological Film Center, 1965–1980," *Journal of the History of the Behavioral Science* 52, no. 4 (2016): 371–91. See also Brian Hochman, *Savage Preservation: The Ethnographic Origins of Modern Media Technology* (Minneapolis: University of Minnesota Press, 2014).

97. Matti Bunzl, "Anthropology beyond Crisis: Toward an Intellectual History of the Extended Present," *Anthropology and Humanism* 30, no. 2 (2005): 187–95. On Project Camelot, see David Price, *Cold War Anthropology: The CIA, the Pentagon, and the Growth of Dual Use Anthropology* (Durham, NC: Duke University Press, 2016), 258–62.

98. Jay Ruby, "Is an Ethnographic Film a Filmic Ethnography?," *Studies in the Anthropology of Visual Communication* 2 (1975): 104–10, on 104. See also Grimshaw, "Visual Anthropology"; Anna Grimshaw, *The Ethnographer's Eye: Ways of Seeing in Modern Anthropology* (Cambridge: Cambridge University Press, 2001); Anna Grimshaw and Amanda Ravetz, *Observational Cinema: Anthropology, Film, and the Exploration of Social Life* (Bloomington: Indiana University Press, 2009).

99. See Simon Schaffer's article on this phenomenon: "'On Seeing Me Write': Inscription Devices in the South Seas," *Representations* 97, no. 1 (2007): 90–122.

100. Claude Lévi-Strauss, "A Writing Lesson," in *Tristes Tropiques*, trans. John Weightman and Doreen Weightman (1955; New York: Penguin Books, 2012), chap. 28.

101. Claude Lévi-Strauss, "Anthropology: Its Achievements and Future," *Current Anthropology* 7 (1966): 124–27, on 127.

102. For a reading of this exchange, see Edward Baring, *The Young Derrida and French Philosophy, 1945–1968* (Cambridge: Cambridge University Press, 2011), chap. 8.

103. Jacques Derrida, *Of Grammatology*, trans. Gayatri Spivak (1967; Baltimore: Johns Hopkins University Press, 1976), 119, 118.

104. "Arche-writing" is a central term developed in Derrida, *Of Grammatology*.

105. Lévi-Strauss, "Writing Lesson," 290.

Chapter 7: Diseases of the Body Image and the Ambiguous Mirror

1. Tanja Legenbauer and Silja Vocks, *Manual der kognitiven Verhaltenstherapie bei Anorexie und Bulimie* (Berlin: Springer, 2006), 266 (ellipses in original).

2. The English physician Richard Morton, quoted in Joan Brumberg, *Fasting Girls: The History of Anorexia Nervosa*, rev. ed. (New York: Vintage, 2000), 46.

3. Randy Schmidt, *Little Girl Blue: The Life of Karen Carpenter* (Chicago: Chicago Review Press, 2010).

4. See, e.g., *People* magazine, November 21, 1983.

5. Brumberg, *Fasting Girls*, 7. Other historians challenged this interpretation, instead asserting the historical specificity of fasting in different times. See esp. Caroline Walker Bynum, *Holy Feast and Holy Fast: The Religious Significance of Food in Medieval Women* (Berkeley: University of California Press, 1987). Brumberg herself is careful not to assert a simple continuity between medieval and modern food refusal, between "anorexia mirabilis" and "anorexia nervosa." See, e.g., Brumberg, *Fasting Girls*, 44.

6. Hilde Bruch, *The Golden Cage: The Enigma of Anorexia Nervosa* (1978; Cambridge, MA: Harvard University Press, 2001), 5.

7. H. Bruch, *Golden Cage*, 3.

8. Hilde Bruch moved away from the body image in her later work, toward a more general perception of a deficient self. See, e.g., Hilde Bruch, *Conversations with Anorexics*, ed. Danita Czyzewiski and Melanie A. Suhr (1988; Northvale, NJ: Jason Aronson 1994), 4–5. This, however, did not entail a change in her therapeutic recommendations of supporting the anorexic patient "in her search for autonomy and self-directed identity by evoking awareness of impulses, feelings, and needs that originate within herself" (8).

9. *Diagnostic and Statistical Manual of Mental Disorders*, 5th ed. (DSM-5), s.v. "Feeding and Eating Disorders: Anorexia Nervosa," https://doi.org/10.1176/appi.books.9780890425596.dsm10#CIHFFIAE.

10. Henry Head and Gordon Holmes, "Sensory Disturbances from Cerebral Lesions," *Brain* 34, nos. 2–3 (1911): 102–254, on 189.

11. Head and Holmes, "Sensory Disturbances," 188.

12. Paul Schilder, *The Image and Appearance of the Human Body* (1935; New York: International Universities Press, 1950), 11.

13. See Katja Guenther, *Localization and Its Discontents: A Genealogy of Psychoanalysis and the Neuro Disciplines* (Chicago: University of Chicago Press, 2015).

14. Schilder, *Image and Appearance*, 11.

15. Paul Schilder, *Das Körperschema: Ein Beitrag zur Lehre vom Bewusstsein des eigenen Körpers* (Berlin: Springer, 1923), 2–3.

16. Hilde Bruch, *Eating Disorders: Obesity, Anorexia Nervosa, and the Person Within* (New York: Basic Books, 1973), 89. See also J. R. Smythies, "The Experience and Description of the Human Body," *Brain* 76 (1953): 132–45. For histories of the body image, see esp. L. C. Kolb, "Disturbances of Body-Image," in *American Handbook of Psychiatry*, ed. Silvano Arieti, vol. 1 (New York: Basic Books, 1959); see also Seymour Fisher and Sidney E. Cleveland, *Body Image and Personality* (Princeton, NJ: Van Nostrand, 1958); Josef Gerstmann, "Psychological and Phenomenological Aspects of Disorders of Body Image," *Journal of Nervous and Mental Disease* 126, no. 6 (1958): 499–512; Douwe Tiemersma, *Body Schema and Body Image: An Interdisciplinary and Philosophical Study* (Amsterdam: Swets en Zeitlinger, 1989).

17. The sponsors of the symposium were the Department of Psychiatry at the Baylor University College of Medicine, the Houston State Psychiatric Institute, and the Veterans Administration Hospital in Houston, Texas. Papers of Hilde Bruch, Manuscript Collection No. 7 of the John P. McGovern Historical Collections and Research Center, Houston Academy of Medicine, Texas Medical Center Library, MS007 Bruch SV, box 1, folder 25, Research in Body Image Symposium, 1965.

18. Papers of Hilde Bruch, MS007 Bruch SV box 1, folder 25, body image note cards.

19. Hilde Bruch, "Perceptual and Conceptual Disturbances in Anorexia Nervosa," *Psychosomatic Medicine* 24, no. 2 (1962): 187–94, on 188.

20. H. Bruch, "Perceptual and Conceptual Disturbances," 189.

21. H. Bruch, *Eating Disorders*, 88. Bruch worked and wrote at a time when mother blaming was a widespread practice in medicine and particularly in psychodynamic schools, as some commentators have insisted. Books such as Bruno Bettelheim's *Empty Fortress* (New York: Free Press, 1967) accused mothers of giving autism to their children and most often sons, while the "schizophrenogenic mother" smothered her children into madness. On the possibilities and pathologies of parental love, see esp. Deborah Weinstein, *The Pathological Family: Postwar America and the Rise of Family Therapy* (Ithaca, NY: Cornell University Press, 2013); see also Edward Dolnick, *Madness on the Couch: Blaming the Victim in the Heyday of Psychoanalysis* (New York: Simon and Schuster, 1998); Marga Vicedo, *The Nature and Nurture of Love: From Imprinting to Attachment in Cold War America* (Chicago: University of Chicago Press, 2013); Anne Harrington, "Mother Love and Mental Illness: An Emotional History," *Osiris* 31 (2016): 94–115.

22. Another important part of the background was contemporary developments in child psychology that revised the view of early infancy as a time of dependence on the environment and that emphasized the child as an active shaper of his or her environment. For a history of this view, see Felix Rietmann, "The Mind of the Child" (PhD diss., Princeton University, 2019).

23. Susanne Gallo, "Short Biographies of Mabel Giddings Wilkin, MD, and Hilde Bruch, MD," *Texas Medicine* 90, no. 10 (October 1994): 60–70.

24. Hilde Bruch, "Fromm-Reichmann, Frieda (1889–1957)," in *International Encyclopedia of Psychiatry, Psychology, Psychoanalysis and Neurology*, ed. Benjamin B. Wolman, vol. 5 (New York: Aesculapius, 1977), 140–41. Bruch also mentioned the influence of another neo-Freudian, Harry Stack-Sullivan, on her thinking. Hilde Bruch, "The Constructive Use of Ignorance," in *Explorations in Child Psychiatry*, ed. James E. Anthony (New York: Plenum, 1975), 247–64, on 255; see also Gallo, "Short Biographies," 64; Joanne Hatch Bruch, *Unlocking the Golden Cage: An Intimate Biography of Hilde Bruch, M.D.* (Carlsbad, CA: Gürze Books, 1996), 190.

25. J. Bruch, *Unlocking the Golden Cage*, 163–64.

26. H. Bruch, "Constructive Use of Ignorance," 255.

27. There were of course others who promoted the importance of child development, e.g., Jean Piaget.

28. H. Bruch, "Perceptual and Conceptual Disturbances," 193.

29. Hilde Bruch, "Falsification of Bodily Needs and Body Concept in Schizophrenia," *Archives of General Psychiatry* 6, no. 1 (1962): 18–24, 20.

30. H. Bruch, "Perceptual and Conceptual Disturbances," 192–93.

31. H. Bruch, "Perceptual and Conceptual Disturbances," 193. See also H. Bruch, "Falsification."

32. To be fair, fathers could be perpetrators as well, though this seemed to be the case less often.

33. H. Bruch, *Golden Cage*, 41, 40.

34. H. Bruch, *Eating Disorders*, 56.

35. H. Bruch, *Golden Cage*, 48.

36. Although it needs to be qualified that this happened during a toxic hunger state; the feeling disappeared when the girl had regained weight.

37. H. Bruch, *Eating Disorders*, 252.

38. H. Bruch, *Eating Disorders*, 51.

39. See Hilde Bruch, "Hunger and Instinct" *Journal of Nervous and Mental Disease* 149 (1969): 91–114.

40. H. Bruch, *Golden Cage*, 45.

41. H. Bruch, *Golden Cage*, 46.

42. H. Bruch, *Golden Cage*, 4.

43. H. Bruch, *Golden Cage*, 5.

44. H. Bruch, *Eating Disorders*, 252.

45. H. Bruch, "Perceptual and Conceptual Disturbances," 194.

46. H. Bruch, "Perceptual and Conceptual Disturbances," 194.

47. H. Bruch, "Perceptual and Conceptual Disturbances," 194.

48. H. Bruch, *Golden Cage*, 102.

49. H. Bruch, *Golden Cage*, 102.

50. See also Hilde Bruch, "Perils of Behavior Modification in Treatment of Anorexia Nervosa," *Journal of the American Medical Association* 230 (1974): 1419–22.

51. H. Bruch, *Golden Cage*, 97, 98ff. Often, restoration of weight was achieved through various forms of forced feeding including tube feeding and intravenous hyperalimentation. If the patient was willing to gain weight but unwilling to eat, she could also take a nutritional preparation in fluid form.

52. H. Bruch, *Eating Disorders*, 343.

53. H. Bruch, *Golden Cage*, 117.

54. H. Bruch, *Golden Cage*, 117.

55. Hilde Bruch, "Must Mothers Be Martyrs?," 1952. (It is unclear where Bruch gave the talk; the archives hold only her typescript of the talk.) Papers of Hilde Bruch, MS070 Bruch S3b1f26, on 2. The talk was part of her larger project of educating parents; see her book *Don't Be Afraid of Your Child: A Guide for Perplexed Parents* (New York: Farrar, Straus, and Young, 1952).

56. Papers of Hilde Bruch, MS070 Bruch S3b1f26.

57. Cf. her discussion in her later book *Eating Disorders*, which mostly avoided social critique.

58. H. Bruch, *Eating Disorders*, 285. This view was common in the 1970s, when authors writing about male anorexia rarely referred to more than one case. Nevertheless, male anorexia was taken seriously by many, and Bruch herself devoted an entire chapter to the topic as well as a separate article: H. Bruch, *Eating Disorders*, chap. 15; Hilde Bruch, "Anorexia Nervosa in the Male," *Psychosomatic Medicine* 33 (1971): 31–47. According to the National Eating Disorders Association (NEDA), about one in three patients with eating disorders today is male. https://www.nationaleatingdisorders.org/statistics-research-eating-disorders.

59. H. Bruch, *Golden Cage*, xx. See also H. Bruch, *Eating Disorders*, 99.

60. H. Bruch, *Golden Cage*, viii.

61. H. Bruch, *Eating Disorders*, xx. Her discussion of gender remained at a very general level, and she did not cite her feminist critics. In her foreword to a later edition of the work, the clinical psychologist Catherine Steiner-Adair pointed out Bruch's "inability to think critically" about the role of fashion ideals in the creation of eating disorders. Catherine Steiner-Adair, foreword to *The Golden Cage*, by Hilde Bruch, new ed. (Cambridge, MA: Harvard University Press, 2001), vii–xvii, on xi.

62. At least this is the case for "genuine" anorexia. The category of the "thin fat" who are constantly dieting to keep their weight low have a more culture-bound etiology. H. Bruch, *Eating Disorders*, 198.

63. See, e.g., H. Bruch, "Perceptual and Conceptual Disturbances."

64. Kim Chernin, *The Hungry Self: Women, Eating, and Identity* (New York: Times Books, 1985); Susie Orbach, *Hunger Strike: The Anorectic's Struggle as a Metaphor for Our Age* (London: Faber and Faber, 1986); Susie Orbach, *Fat Is a Feminist Issue: The Anti-diet Guide to Permanent Weight-Loss* (New York: Galahad Books, 1978); Mara Selvini Palazzoli, *Self-Starvation: From Individual to Family Therapy in the Treatment of Anorexia Nervosa*, trans. Arnold Pomerans (1973; New York: Jason Aronson, 1974). There were other feminist accounts that did not neatly fit into this classification; these include those directly engaging with a psychoanalytic point of view, e.g., Marlene Boskind-Lodahl, "Cinderella's Stepsisters: A Feminist Perspective on Anorexia Nervosa and Bulimia," *Signs: Journal of Women in Culture and Society* 2, no. 2 (1976): 342–56.

65. Chernin, *Hungry Self*, 49.

66. Chernin, *Hungry Self*, 136.

67. Chernin, *Hungry Self*, 17.

68. See also Kim Chernin, *Womansize: The Tyranny of Slenderness* (London: Women's, 1983).

69. Orbach, *Fat Is a Feminist Issue*, 175.

70. Orbach, *Fat Is a Feminist Issue*, 178.

71. Orbach, *Fat is a Feminist Issue*, 5.

72. See, for example, Naomi Wolf, *Fire with Fire: The New Female Power and How It Will Change the 21st Century* (New York: Random House, 1993).

73. Naomi Wolf, *The Beauty Myth: How Images of Beauty are Used against Women* (1990; New York: Perennial, 2002), 198.

74. Wolf, *Beauty Myth*, 189.

75. One stone (the imperial unit of mass still commonly used for body weight in the United Kingdom) is equal to 14 pounds or approximately 6.35 kilograms.

76. Wolf, *Beauty Myth*, 186.

77. Sandra A. Birtchnell, J. Hubert Lacey, and Anne Harte, "Body Image Distortion in Bulimia Nervosa," *British Journal of Psychiatry* 147 (1985): 408–12; Mary E. Willmuth et al., "Body Size Distortion in Bulimia Nervosa," *International Journal of Eating Disorders* 4 (1985): 71–78. Overestimation of body size was also found in the obese: Albert J. Stunkart and Myer Mendelson, "Obesity and the Body Image: I. Characteristics of Disturbances in the Body Image of Some Obese Persons," *American Journal of Psychiatry* 123 (1967): 1296–300.

78. Sandra A. Birtchnell, Bridget M. Dolan, and J. Hubert Lacey, "Body Image Distortion in Non–Eating Disordered Women," *International Journal of Eating Disorders* 6, no. 3 (1987): 385–91. In the study by Birtchnell, Dolan, and Lacy, body image was measured in a population of non-eating-disordered women at normal body weight, using a visual size-estimation apparatus. In an experimental design similar to the Askevold line-drawing test (discussed below), subjects were "asked to move two lights on a bar until the distance between them corresponded as accurately as they could judge to their actual body widths at waist, chest, and hip level" (386). They were also asked to estimate the width of a neutral object by the same procedure. The researchers found that, while the neutral object was assessed correctly by the majority of participants, all subjects (from both the eating-disordered and the non-eating-disordered groups) overestimated their own body size. Other studies showed similar results. The visual size-estimation apparatus was first described by P. D. Slade and G.F.M. Russell in 1973: "Experimental Investigations of Bodily Perception in Anorexia Nervosa and Obesity," *Psychotherapy and Psychosomatics* 22 (1973): 359–63.

79. Susan Bordo, *Unbearable Weight: Feminism, Western Culture, and the Body*, new ed. (1993; Berkeley: University of California Press, 2003), xxi.

80. Bordo, *Unbearable Weight*, 55.

81. Bordo, *Unbearable Weight*, 57.

82. Nathan G. Hale, *The Rise and Crisis of Psychoanalysis in the United States: Freud and the Americans, 1917–1985* (New York: Oxford University Press, 1995); John Burnham, ed., *After Freud Left: A Century of Psychoanalysis in America* (Chicago: University of Chicago Press, 2012).

83. Karl Popper, *The Logic of Scientific Discovery* (New York: Basic Books, 1959).

84. Hans J. Eysenck, "The Effects of Psychotherapy: An Evaluation," *Journal of Consulting Psychology* 16 (1952): 319–24, on 323. See also Hans J. Eysenck, *Behavior Therapy and the Neuroses* (Oxford: Pergamon, 1960).

85. There was also considerable overlap and collaboration between behaviorism and psychoanalysis. See Rebecca Lemov, *World as Laboratory: Experiments with Mice, Mazes, and Men* (New York: Hill and Wang, 2005), part 2.

86. Joseph Wolpe, *Psychotherapy by Reciprocal Inhibition* (Stanford, CA: Stanford University Press, 1958). See also Joseph Wolpe and Vivienne C. Rowan, "Panic Disorder: A Product of Classical Conditioning," *Behaviour Research and Therapy* 26, no. 6 (1988): 441–50.

87. See the work of David M. Clark in the UK and D. H. Barlow in the United States, discussed in S. Rachman, "The Evolution of Cognitive Behavior Therapy," in *Science and Practice of Cognitive Behaviour Therapy*, ed. David M. Clark and Christopher G. Fairburn (Oxford: Oxford University Press, 1997), 1–26.

88. The wave metaphor is contested. See Rachman, "Evolution."

89. See, e.g., Katherine Phillips and Raymond Dufresne Jr., "Body Dysmorphic Disorder: A Guide for Primary Care Physicians," *Primary Care* 29, no. 1 (2002): 99–111.

90. Because there was no longer an analyst required who interpreted the patients' manifest symptoms to arrive at their deeper meaning, the CBT approach invited the patients to take their improvement in their own hands. Scholars have noted the affinity between CBT and today's media-based self-help approaches (at least those targeting a particular disorder); there are many different such approaches, from group based (Alcoholics Anonymous is the epitome) to media based (e.g., "bibliotherapy"). This is not surprising as they share with CBT the use of a structured protocol. See Patti Lou Watkins, "Self-Help Therapies: Past and Present," in *Handbook of Self-Help Therapies*, ed. Patti Lou Watkins and George A. Clum (New York: Routledge, 2007), 1–24, on 3.

91. Quoted in H. Bruch, *Eating Disorders*, 90.

92. H. Bruch, *Eating Disorders*, 189.

93. H. Bruch, *Eating Disorders*, 90.

94. Arthur C. Traub and J. Orbach, "Psychophysical Studies of Body-Image: I. The Adjustable Body-Distorting Mirror," *Archives of General Psychiatry* 11 (1964): 53–66, on 54. Traub and Orbach were part of a larger group working on body image distortion procedures. For an overview, see S. Skrzypek, P. M. Wehmeier, and H. Remschmidt, "Body Image Assessment Using Body Size Estimation in Recent Studies on Anorexia Nervosa: A Brief Review," *European Child and Adolescent Psychiatry* 10 (2001): 215–21. It is noteworthy that they distinguish between body part (mostly estimation) and whole body (mostly distortion) procedures. For a book-length study of body perception, see Franklin C. Shontz, *Perceptual and Cognitive Aspects of Body Experience* (New York: Academic, 1969).

95. Traub and Orbach, "Psychophysical Studies of Body-Image," 60–61. The authors focused on vision for heuristic reasons: "Despite obvious interaction between sensory modalities, there is an heuristic benefit in considering the contributions of each separately" (54).

96. Traub and Orbach, "Psychophysical Studies of Body-Image," 61.

97. Note, however, that Traub and Orbach also acknowledged the body image to be plastic. As part of their results, they reported: "After some minutes of adjustment, many subjects declare, often quite sheepishly, that they have forgotten precisely what they look like. Some subjects urgently request that they be permitted to examine themselves in a normal mirror before they proceed." Traub and Orbach, "Psychophysical Studies of Body-Image," 65.

98. Finn Askevold, "Measuring Body Image: Preliminary Report on a New Method," *Psychotherapy and Psychosomatics* 26 (1975): 71–77, on 71.

99. Askevold, "Measuring Body Image," 76.

100. See also Slade and Russell, "Experimental Investigations."

101. Mirror treatment is of course only one part of CBT for eating disorders. For a fuller account, see David Garner and Kelly Bemis, "A Cognitive-Behavioral Approach to Anorexia Nervosa," *Cognitive Therapy and Research* 6, no. 2 (1982): 123–50.

102. Although the emotional resistance to the mirror needed to be overcome to begin the treatment, the strong emotional reaction was also considered a good thing because it made the treatment more effective. Adrienne Key et al., "Body Image Treatment within an Inpatient Program for Anorexia Nervosa: The Role of Mirror Exposure in the Desensitization Process," *International Journal of Eating Disorders* 31, no. 2 (2002): 185–90.

103. Thomas F. Cash and Jill R. Grant, "The Cognitive-Behavioural Treatment of Body-Image Disturbances," in *Sourcebook of Psychological Treatment Manuals for Adult Disorders*, ed.

Vincent B. Van Hasselt and Michel Hessen (New York: Plenum, 1996), 567–614, on 593. See also Thomas F. Cash's self-help book *What Do You See When You Look in the Mirror? Helping Yourself to a Positive Body Image* (New York: Bantam, 1995), chap. 3.

104. Cash and Grant, "Cognitive-Behavioural Treatment," 593.

105. Legenbauer and Vocks, *Manual der kognitiven Verhaltenstherapie*, 250.

106. Legenbauer and Vocks, *Manual der kognitiven Verhaltenstherapie*, 250.

107. Legenbauer and Vocks, *Manual der kognitiven Verhaltenstherapie*, 250.

108. Legenbauer and Vocks, *Manual der kognitiven Verhaltenstherapie*, 252.

109. Silja Vocks and Tanja Legenbauer, *Körperbildtherapie bei Anorexia und Bulimia Nervosa: Ein kognitiv-verhaltenstherapeutisches Behandlungsprogramm* (Göttingen: Hogrefe, 2005), 96. The wings are not a universal recommendation, but almost all papers emphasize the sufficient size of the mirror. The manual moves back and forth between the female and male pronoun. Part of the reason is that in the German language, despite the linguistic political action of the past decades, the male pronoun is still used to stand in for both genders, thus allowing a certain indeterminacy, which the authors exploit.

110. See, e.g., Cash and Grant, "Cognitive-Behavioural Treatment."

111. Vocks and Legenbauer, "Körperkonfrontation," chap. 9 in *Körperbildtherapie bei Anorexia und Bulimia Nervosa*, 92–113, on 97.

112. In a variation of the procedure, the patient was videotaped during her confrontation exercise so that later on, when viewing the video with the therapist, she could compare the image of herself in the mirror with her descriptions. Vocks and Legenbauer, "Körperkonfrontation," 102.

113. See, e.g., Garner and Bemis, "Cognitive-Behavioral Approach to Anorexia Nervosa."

114. Vocks and Legenbauer, *Körperbildtherapie*, 269.

115. Vocks and Legenbauer, *Körperbildtherapie*, 269.

116. Vocks and Legenbauer, *Körperbildtherapie*, 269.

117. Vocks and Legenbauer, *Körperbildtherapie*, 269.

Chapter 8: Imperfect Reflections

1. The Institute of Human Physiology became the Department of Neuroscience in 2004 and is now part of the Medical and Surgery Department at the University of Parma.

2. Giacomo Rizzolatti et al., "Premotor Cortex and the Recognition of Motor Actions," *Cognitive Brain Research* 3 (1996): 131–41. The term *mirror neuron* subsequently appeared again in the follow-up paper: Vittorio Gallese et al., "Action Recognition in the Premotor Cortex," *Brain* 119 (1996): 593–609. The class of premotor neurons that showed the mirroring phenomenon was first described in G. Di Pellegrino et al., "Understanding Motor Events: A Neurophysiological Study," *Experimental Brain Research* 91 (1992): 176–80.

3. Among Rizzolatti's PhD and postdoctoral students were Leonardo Fogassi, Anna Berti, Carlo Porro, Vittorio Gallese, Luciano Fadiga, and Christian Keysers.

4. G. Rizzolatti et al., "Functional Organization of Inferior Area 6 in the Macaque Monkey: II. Area F5 and the Control of Distal Movements," *Experimental Brain Research* 71 (1988): 491–507, on 503.

5. Wilder Penfield and Theodore Rasmussen, *The Cerebral Cortex of Man: A Clinical Study of Localization of Function* (New York: Macmillan, 1950).

6. C. N. Woolsey et al., "Patterns of Localization in Precentral and 'Supplementary' Motor Areas and Their Relation to the Concept of a Premotor Area," *Research Publications—Association for Research in Nervous and Mental Disease* 30 (1952): 238–64. Woolsey's "simiusculus" was based on Penfield's representation. The medical historian Erwin Ackerknecht, Woolsey's colleague at the University of Wisconsin, suggested to him this simian equivalent. Four years earlier, Woolsey had mapped the tactile areas of the rat's cerebral cortex. C. N. Woolsey and D. H. Le Messurier, "The Pattern of Cutaneous Representation in the Rat's Cerebral Cortex," *Federation Proceedings* 7 (1948): 137.

7. For a history of this approach within the larger tradition of localization of function, see Katja Guenther, *Localization and Its Discontents: A Genealogy of Psychoanalysis and the Neuro Disciplines* (Chicago: University of Chicago Press, 2015).

8. Giacomo Rizzolatti and Corrado Sinigaglia, *Mirrors in the Brain: How Our Minds Share Actions and Emotions*, trans. Frances Anderson (Oxford: Oxford University Press, 2008), 3.

9. A note on terminology: For Penfield and Woolsey, area 4 was the "primary motor cortex," and area 6 a region subordinate to it; they rejected the notion of a functionally separate "premotor cortex." Rizzolatti and colleagues argued against this view in their attempt to grant the premotor cortex (area 6) a function independent of area 4.

10. M. Gentilucci et al., "Functional Organization of Inferior Area 6 in the Macaque Monkey: I. Somatotopy and the Control of Proximal Movements," *Experimental Brain Research* 71 (1988): 475–90, on 476. The paper is the first part of a pair: G. Rizzolatti et al., "Functional Organization."

11. Gentilucci et al., "Functional Organization," 476.

12. Rizzolatti et al., "Functional Organization," 506.

13. Gentilucci et al., "Functional Organization," 481, 485.

14. Rizzolatti et al., "Functional Organization," 506.

15. Gentilucci et al., "Functional Organization," 487.

16. Gentilucci et al., "Functional Organization," 487.

17. G. Rizzolatti et al., "Neurons Related to Reaching-Grasping Arm Movements in the Rostral Part of Area 6 (Area 6aβ)," *Experimental Brain Research* 82 (1990): 337–50.

18. In order to make their complexity argument, the researchers needed to set aside the possibility that these were "set-related" neurons. Set-related neurons had been found in various motor areas and were believed to reflect "the motor aspect of the movement preparation rather than sensory or motivational factors." Rizzolatti et al., "Neurons Related," 347.

19. Di Pellegrino et al., "Understanding Motor Events," 176.

20. Di Pellegrino et al., "Understanding Motor Events," 177.

21. Rizzolatti et al., "Premotor Cortex," 135. See also Gallese et al., "Action Recognition," 600–602.

22. Gallese et al., "Action Recognition," 601–2.

23. Gallese et al., "Action Recognition," 601.

24. Rizzolatti et al., "Premotor Cortex," 135.

25. Ironically, the shift in the experiment and the emergence of mirroring as an explicit element of the discussion made the one-way mirror superfluous. If the stimulus was the movement of the scientist rather than a piece of food, it no longer made sense to stash it behind a piece of glass.

26. On the hype around mirror neurons, see, e.g., Gregory Hickok, *The Myth of Mirror Neurons: The Real Neuroscience of Communication and Cognition* (New York: Norton, 2014). On the recent

decline of interest in mirror neurons, see Cecilia Heyes and Caroline Catmur, "What Happened to Mirror Neurons?," *Perspectives on Psychological Science* 17, no. 1 (2022): 153–68.

27. Bruno Wicker et al., "Both of Us Disgusted in *My* Insula: The Common Neural Basis of Seeing and Feeling Disgust," *Neuron* 40 (2003): 655–64, on 660.

28. Wicker et al., "Both of Us Disgusted," 660.

29. Wicker et al., "Both of Us Disgusted," 661.

30. Wicker et al., "Both of Us Disgusted," 661.

31. Wicker's assumption of "basic emotions" can be traced back to the work of American psychologists Silvan S. Tomkins and Paul Ekman. Ruth Leys, "'Both of Us Disgusted in *My* Insula': Mirror-Neuron Theory and Emotional Empathy," in *Science and Emotions after 1945: A Transatlantic Perspective*, ed. Frank Biess and Daniel M. Gross (Chicago: University of Chicago Press, 2014): 67–95, on 73.

32. TT is often lumped together with the broader "theory of mind" (ToM) to explain the understanding of the outside world, and in particular the mental states of others. Here, I use "theory of mind" in a broader sense: it can be either ST or TT. For a short history of TT, see Marco Iacoboni, *Mirroring People: The Science of Empathy and How We Connect with Others* (New York: Farrar, Straus and Giroux, 2008), 168–69.

33. Vittorio Gallese and Alvin Goldman, "Mirror Neurons and Simulation Theory of Mind-Reading," *Trends in Cognitive Science* 2, no. 12 (1998): 493–501, on 493. Another term for *naive theory* is *folk* or *commonsense psychology*, which refers to the ability to understand mental states in others. Theory theory was first formulated in the 1980s within developmental psychology: Margaret Boden, *Mind as Machine: A History of Cognitive Science* (Oxford: Oxford University Press, 2006).

34. J.H.G. Williams et al., "Imitation, Mirror Neurons and Autism," *Neuroscience and Biobehavioral Reviews* 25 (2001): 287–95, on 288.

35. Vittorio Gallese, "The 'Shared Manifold' Hypothesis: From Mirror Neurons to Empathy," *Journal of Consciousness Studies* 8 (2001): 33–50, on 45 (my emphasis).

36. The key papers are Leys, "'Both of Us Disgusted'" (for her larger account of the history of emotions, see Ruth Leys, *The Ascent of Affect: Genealogy and Critique* [Chicago: University of Chicago Press, 2017]); Allan Young, "The Social Brain and the Myth of Empathy," *Science in Context* 25, no. 3 (2012): 401–24; Allan Young, "Mirror Neurons and the Rationality Problem," in *Rational Animals, Irrational Humans*, ed. S. Watanabe, A. P. Blaisdell, L. Huber, A. Young (Tokyo: Keio University Press, 2009), 67–80. Susan Lanzoni has, among other things, shown how mirror neuron research draws on pragmatism and phenomenology in support of its theories of empathy: Susan Lanzoni, "Imagining and Imaging the Social Brain: The Case of Mirror Neurons," *Canadian Bulletin of Medical History* 33, no. 2 (2016): 447–64. For a larger history of empathy, and the place of mirror neurons within it, see Susan Lanzoni, *Empathy: A History* (New Haven, CT: Yale University Press, 2018), esp. chap. 9.

37. Leys, "'Both of Us Disgusted,'" 72. For an important extension of this argument, see Ruth Leys, *Newborn Imitation: The Stakes of a Controversy* (Cambridge, UK: Cambridge University Press, 2020), esp. 27–37.

38. Leys, "'Both of Us Disgusted.'" Young agrees with Leys in that mirroring is depicted as a direct process, a "brain-to-brain product, unmediated by mental states." Young, "Social Brain," 403.

39. I pair autism and psychopathy here only because that opposition was important for the scientists I study.

40. Harma Meffert et al., "Reduced Spontaneous but Relatively Normal Deliberate Vicarious Representations in Psychopathy," *Brain* 136 (2013): 2550–62.

41. Christian Keysers, *The Empathic Brain: How the Discovery of Mirror Neurons Changes Our Understanding of Human Nature* (n.p.: Social Brain, 2011), 205–19.

42. Keysers, *Empathic Brain*, 206.

43. Their diagnosis was made according to DSM-4 criteria. Meffert et al., "Reduced Spontaneous."

44. Meffert et al., "Reduced Spontaneous," 2558. That is, scanned in a Philips Intera 3 T Quasar: 2559, 2553.

45. Christian Keysers, "Inside the Mind of a Psychopath—Empathic, but Not Always," *Psychology Today* (blog), July 24, 2013, http://www.psychologytoday.com/blog/the-empathic-brain/201307/inside-the-mind-psychopath-empathic-not-always.

46. Keysers, "Inside the Mind of a Psychopath."

47. The avoidance of the term *mirror neurons* is consistent with the decline of interest in mirror neuron research after the "period of most intense interest between 2002 and 2007." Young, "Social Brain," 403.

48. Meffert et al., "Reduced Spontaneous," 2552.

49. Meffert et al., "Reduced Spontaneous," 2552.

50. Meffert et al., "Reduced Spontaneous," 2559.

51. Keysers, "Inside the Mind of a Psychopath." To the researchers, this suggested possibilities for treatment. Meffert et al., "Reduced Spontaneous," 2560.

52. Meffert et al., "Reduced Spontaneous," 2560.

53. Shirley Fecteau, Alvaro Pascual Leone, and Hugo Théoret, "Psychopathy and the Mirror Neuron System: Preliminary Findings from a Non-psychiatric Sample," *Psychiatry Research* 160 (2008): 137–44, on 143.

54. Fecteau, Pascual-Leone, and Théoret, "Psychopathy," 142. A. Avenanti, D. Bueti, G. Galati, S. M. Aglioti, "Transcranial Magnetic Stimulation Highlights the Sensorimotor Side of Empathy for Pain," *Nature Neuroscience* 8, no. 7 (2005): 955–60. See also the earlier discussion in the paper of distinctions between forms of empathy, such as R. J. Blair's distinction between cognitive, motor, and emotional empathy (143). R. J. Blair, "Responding to the Emotions of Others: Dissociating Forms of Empathy through the Study of Typical and Psychiatric Populations," *Consciousness and Cognition* 14, no. 4 (2005): 698–718.

55. Fecteau, Pascual-Leone, and Théoret, "Psychopathy," 142.

56. Fecteau, Pascual-Leone, and Théoret, "Psychopathy," 142.

57. Lindsay M. Oberman et al., "EEG Evidence for Mirror Neuron Dysfunction in Autism Spectrum Disorders," *Cognitive Brain Research* 24 (2005): 190–98; Justin H. G. Williams et al., "Neural Mechanisms of Imitation and 'Mirror Neuron' Functioning in Autistic Spectrum Disorder," *Neuropsychologia* 44 (2006): 610–21. V. S. Ramachandran's group had first presented results from EEG experiments at the annual meeting of the Society for Neuroscience in 2000. E. L. Altschuler et al., "Mu Wave Blocking by Observation of Movement and Its Possible Use to Study the Theory of Other Minds," annual meeting of the Society for Neuroscience (2000), abstract 68.1.

58. Ramachandran's more speculative papers were usually not coauthored, which is itself exceptional in the age of scientific collaboration. The lack of connection with the Parma group can be explained by the fact that Ramachandran's team did not share the same research paradigm: instead of the rather elaborate electrophysiological research in monkeys that the Italians used, Ramachandran's group did research in humans with lower-tech apparatuses, such as EEG and behavioral studies.

59. Oberman et al., "EEG Evidence," 195.

60. V. S. Ramachandran and Lindsay M. Oberman, "Broken Mirrors," *Scientific American*, November 2006, 63–69.

61. Williams et al., "Neural Mechanisms." The Iacoboni paper is Marco Iacoboni et al., "Cortical Mechanisms of Human Imitation," *Science* 286 (1999): 2526–28.

62. Williams et al., "Neural Mechanisms," 620.

63. Williams et al., "Neural Mechanisms," 618.

64. The subjects were asked to imitate lip forms seen presented to them in still pictures. Nobuyuki Nishitani, Sari Avikainen, and Riitta Hari, "Abnormal Imitation-Related Cortical Activation Sequences in Asperger's Syndrome," *Annals of Neurology* 55 (2004): 558–62.

65. See, e.g., Simon Baron-Cohen and Uta Frith on theory of mind, Alyson Bacon on empathy, and Margaret Kjelgaard and Helen Tager-Flusburg on language. Simon Baron-Cohen, "Theory of Mind and Autism: A Review," in *International Review of Research in Mental Retardation: Autism*, vol. 23, ed. Laraine Masters Glidden (San Diego: Academic, 2001), 169–84; Uta Frith, *Autism: Explaining the Enigma* (Oxford: Basil Blackwell, 1989); Alyson L. Bacon et al., "The Responses of Autistic Children to the Distress of Others," *Journal of Autism and Developmental Disorders* 28 (1998): 129–42; Margaret M. Kjelgaard and Helen Tager-Flusburg, "An Investigation of Language Impairment in Autism: Implications for Genetic Subgroups," *Language and Cognitive Processes* 16 (2001): 287–308.

66. Williams et al., "Imitation, Mirror Neurons and Autism."

67. Williams et al., "Imitation, Mirror Neurons and Autism," 289.

68. Williams et al., "Imitation, Mirror Neurons and Autism," 291.

69. Paul D. McGeoch, David Brang, and V. S. Ramachandran, "Apraxia, Metaphor and Mirror Neurons," *Medical Hypotheses* 69 (2007): 1165–68. David Brang, who is now at Northwestern University, was at the time a doctoral student of Ramachandran's.

70. McGeoch, Brang, and Ramachandran, "Apraxia, Metaphor and Mirror Neurons" 1167.

71. Oberman et al., "EEG Evidence," 196.

72. Note however that some suggested a general flattening of emotions in psychopathic individuals.

73. Fecteau, Pascual-Leone, and Théoret, "Psychopathy," 142.

74. But alas, the relationship was not always as clear. In a letter to the editor in response to Meffert et al.'s paper, Oberman and colleagues pointed out that in autism as well, there was a "dissociation between spontaneous and deliberate experience." For example, individuals with Asperger's syndrome were able to "understand mental states such as desires and beliefs (mentalizing) when explicitly prompted to do so" even though they showed difficulty in doing so spontaneously. Steven M. Gillespie, Joseph P. McCleery, and Lindsay M. Oberman, "Spontaneous versus Deliberate Vicarious Representations: Different Routes to Empathy in Psychopathy and Autism," *Brain* 137 (2014): e272 (1–3), on 1. However, there was a difference in the "route to

empathy" in individuals with autism and those who were psychopathic. Autistic individuals "exhibit relatively more intact social and empathic activations [that is, activations of brain areas involved in the processing of social behavior or emotions] when the observed individuals are socially and emotionally relevant to them, or when they are given explicit instructions to do so" (2). But they showed reduced activations when confronted with individuals unknown to them. Psychopathic individuals, on the other hand, showed reduced spontaneous activation only when dealing with emotional stimuli, not during the social situation; their "social brain regions were active under both spontaneous and deliberate conditions" (2). We also need to keep in mind that mirrors show resemblance as well as inversion.

75. Di Pellegrino et al., "Understanding Motor Events," 176.

76. L. Fadiga et al., "Motor Facilitation during Action Observation: A Magnetic Stimulation Study," *Journal of Neurophysiology* 73, no. 6 (1995): 2608–11. This grew out of a collaboration between Rizzolatti's group at the Istituto di Fisiologia and the Center for Clinical Neurology in Parma. In the study, the researchers measured so-called motor evoked potentials (MEPs) from the hand muscles of their human subjects. The MEPs were triggered through simultaneous application of transcranial magnetic stimulation (TMS) to the motor areas of the brain. Then, in order to explore the "correlation" between observed and executed action, in an additional trial in the same study, they recorded the electromyogram activity of the set of muscles they had explored in the first experiment, while the subjects were executing grasping and other movements. In other words, the researchers compared brain activity measured through MEP and muscular activity measured through electromyography, to compare observation and execution. They found that the MEP pattern "reflected the pattern of muscle activity" in the electromyogram experiment, and that they were "very similar" (2608, 2609).

77. G. Rizzolatti et al., "Localization of Grasp Representations in Humans by PET: 1. Observation versus Execution," *Experimental Brain Research* 111 (1996): 246–52.

78. The first time mirror neurons were described in humans based on actual recordings was in 2010, in an operation on an epileptic patient. Roy Mukamel et al., "Single-Neuron Responses in Humans during Execution and Observation of Actions," *Current Biology* 20 (2010): 750–56.

79. Rizzolatti et al., "Localization of Grasp," 250.

80. Rizzolatti et al., "Localization of Grasp," 249, 250. The situation was less straightforward for area 45: although the area was cytoarchitectonically similar in both species in certain respects, it was not clear that there was a functional homology between monkeys and humans.

81. Rizzolatti et al., "Localization of Grasp," 250.

82. Rizzolatti et al., "Localization of Grasp," 250.

83. V. S. Ramachandran, "The Neurology of Self-Awareness: The *Edge* 10th Anniversary Essay," *Edge* (website), January 8, 2007, http://edge.org/3rd_culture/ramachandran07 /ramachandran07_index.html. See also V. S. Ramachandran, *The Tell-Tale Brain: A Neuroscientist's Quest for What Makes Us Human* (New York: W. W. Norton, 2011).

84. V. S. Ramachandran, "Mirror Neurons and Imitation Learning as the Driving Force behind the Great Leap Forward in Human Evolution," *Edge* (website), May 31, 2000, https:// edge.org/conversation/mirror-neurons-and-imitation-learning-as-the-driving-force-behind -the-great-leap-forward-in-human-evolution/.

85. V. S. Ramachandran, "The Neurons That Shaped Civilization," TED talk, November 2009, https://www.ted.com/talks/vilayanur_ramachandran_the_neurons_that_shaped_civilization.

86. Ramachandran, "Mirror Neurons."

87. Ramachandran, "Mirror Neurons."

88. Ramachandran, "Neurology of Self-Awareness." Both arguments were also put forward in Ramachandran's book *The Tell-Tale Brain*.

89. Iacoboni, *Mirroring People*, 202. Rizzolatti argued along similar lines for "increasingly articulated and complex mirror neuron systems" in humans. Rizzolatti and Sinigaglia, *Mirrors in the Brain*, 192.

90. For this reason, they have also been called "anti-mirror neurons." Christian Keysers and Valeria Gazzola, "Social Neuroscience: Mirror Neurons Recorded in Humans," *Current Biology* 20 (2010): R353–R354.

91. Iacoboni, *Mirroring People*, 193.

92. Mukamel et al., "Single-Neuron Responses," 754.

93. Mukamel et al., "Single-Neuron Responses," 754–55.

94. Giacomo Rizzolatti and Michael A. Arbib, "Language within Our Grasp," *Trends in Neurosciences* 21, no. 5 (1998): 188–94.

95. Rizzolatti and Arbib, "Language within Our Grasp," 190.

96. Rizzolatti and Arbib, "Language within Our Grasp," 190.

97. Rizzolatti and Arbib, "Language within Our Grasp," 191.

98. Rizzolatti and Arbib, "Language within Our Grasp," 193.

99. Rizzolatti and Arbib, "Language within Our Grasp," 193.

100. Maurizio Gentilucci et al., "Grasp with Hand and Mouth: A Kinematic Study on Healthy Subjects," *Journal of Neurophysiology* 86 (2001): 1685–99. In 2003 Gentilucci found similar responses when subjects were merely observing grasping movements. Maurizio Gentilucci, "Grasp Observation Influences Speech Production," *European Journal of Neuroscience* 17 (2003): 179–84.

101. Rizzolatti and Arbib, "Language within Our Grasp," 193.

102. V. S. Ramachandran, "Phantom Limbs, Neglect Syndromes, Repressed Memories, and Freudian Psychology," *International Review of Neurobiology* 37 (1994): 291–333. Ramachandran's phantom limbs resonate, of course, with Edmund Carpenter's phantoms. Marshall McLuhan, who saw electronic media as the extensions of men, came even more closely to the meaning of Ramachandran's phantom limbs, which were projections of the body image into space.

103. V. S. Ramachandran and William Hirstein, "The Perception of Phantom Limbs: The D. O. Hebb Lecture," *Brain* 121 (1998): 1603–30, on 1604.

104. V. S. Ramachandran, "Phantom Limbs," 317.

105. V. S. Ramachandran and Sandra Blakeslee, *Phantoms in the Brain: Probing the Mysteries of the Human Mind* (New York: Harper Perennial, 1998), 46.

106. Ramachandran, "Phantom Limbs," 302; Ramachandran and Hirstein, "Perception of Phantom Limbs," 1621.

107. V. S. Ramachandran and Diane Rogers-Ramachandran, "Synaesthesia in Phantom Limbs Induced with Mirrors," *Proceedings of the Royal Society of London B* 263 (1996): 377–86, on 383.

108. V. S. Ramachandran et al., "Illusions of Body Image: What They Reveal about Human Nature," in *The Mind-Brain Continuum: Sensory Processes*, ed. Rodolfo Llinás and Patricia S. Churchland (Cambridge: MIT Press, 1996), 29–60, on 31.

109. Ramachandran and Rogers-Ramachandran, "Synaesthesia," 381.

110. V. S. Ramachandran and Diane Rogers-Ramachandran, "Sensations Referred to a Patient's Phantom Arm from Another Subject's Intact Arm: Perceptual Correlates of Mirror Neurons," *Medical Hypotheses* 70, no. 6 (2008): 1233–34, on 1233. He first called the action of mirror neurons a "VR (virtual reality) simulation," in a popular article from 2000: Ramachandran, "Mirror Neurons." See also V. S. Ramachandran and David Brain, "Sensations Evoked in Patients with Amputation from Watching an Individual Whose Corresponding Intact Limb Is Being Touched," *Archives of Neurology* 66 (2009): 1281–84, on 1281.

111. V. S. Ramachandran and Eric L. Altschuler, "The Use of Visual Feedback, in Particular Mirror Visual Feedback, in Restoring Brain Function," *Brain* 132 (2009): 1693–710, on 1702.

112. Ramachandran and Altschuler, "Use of Visual Feedback," 1702.

113. "The 2011 Time 100," *Time*, April 21, 2011, http://content.time.com/time/specials/packages/completelist/0,29569,2066367,00.html.

114. See, e.g., Ramachandran and Blakeslee, *Phantoms in the Brain*; Ramachandran, *Tell-Tale Brain*; V. S. Ramachandran, *A Brief Tour of Human Consciousness: From Impostor Poodles to Purple Numbers* (New York: Pi, 2004), 1, 7–8; Ramachandran et al., "Illusions of Body Image"; Ramachandran, "Phantom Limbs."

115. Ramachandran et al., "Illusions of Body Image," 29.

116. Ramachandran et al., "Illusions of Body Image," 30.

117. For a critique of "neuropsychoanalysis," see Nima Bassiri, "Freud and the Matter of the Brain: On the Rearrangements of Neuropsychoanalysis," *Critical Inquiry* 40 (Autumn 2013): 1–26.

118. Ramachandran, "Phantom Limbs," 316.

119. Ramachandran et al., "Illusions of Body Image," 39.

120. Ramachandran et al., "Illusions of Body Image," 39–40.

121. V. S. Ramachandran and Paul D. McGeoch, "Occurrence of Phantom Genitalia after Gender Reassignment Surgery," *Medical Hypotheses* 69, no. 5 (2007): 1001–3. Ramachandran did not refer to Freud or psychoanalysis in this publication, however, presumably because Freud's views on transsexuality, homosexuality, and bisexuality are complex.

122. V. S. Ramachandran and Diane Rogers-Ramachandran, "It's All Done with Mirrors," *Scientific American Mind* 18, no. 4 (August/September 2007): 16–18, on 16.

Conclusion

1. Katja Guenther, *Localization and Its Discontents: A Genealogy of Psychoanalysis and the Neuro Disciplines* (Chicago: University of Chicago Press, 2015).

2. See, e.g., Ian Hacking, *Representing and Intervening: Introductory Topics in the Philosophy of Natural Science* (Cambridge: Cambridge University Press, 1983).

3. These books focus predominantly on philosophical developments though several of them dedicate chapters to science as well: Charles Taylor, *Sources of the Self: The Making of the Modern Identity* (Cambridge, MA: Harvard University Press, 1989); Jerrold Seigel, *The Idea of the Self: Thought and Experience in Western Europe since the Seventeenth Century* (Cambridge: Cambridge University Press, 2005); Gerald Izenberg, *Identity: The Necessity of a Modern Idea* (Philadelphia: University of Pennsylvania Press, 2016); George Makari, *Soul Machine: The Invention of the*

Modern Mind (New York: Norton, 2015); Raymond Marin and John Baresi, *The Rise and Fall of Soul and Self: An Intellectual History of Personal Identity* (New York: Columbia University Press, 2006); Dror Wahman, *The Making of the Modern Self: Identity and Culture in Eighteenth-Century England* (New Haven, CT: Yale University Press, 2004). For histories of the self that touch on psychological discourses, see esp. Jan Goldstein, *The Post-revolutionary Self: Politics and Psyche in France, 1750–1850* (Cambridge, MA: Harvard University Press, 2008); and *Nikolas Rose: Inventing Our Selves: Psychology, Power, and Personhood* (Cambridge: Cambridge University Press, 1998).

4. Fernando Vidal and Francisco Ortega, *Being Brains: Making the Cerebral Subject* (New York: Fordham University Press, 2017).

5. On brain death and the cerebral subject, see esp. Margaret Lock, *Twice Dead: Organ Transplants and the Reinvention of Death* (Berkeley: University of California Press, 2002).

6. On lobotomy, see Jack Pressman, *Last Resort: Psychosurgery and the Limits of Medicine* (Cambridge: Cambridge University Press, 1998); Mical Raz, *The Lobotomy Letters: The Making of American Psychosurgery* (Rochester, NY: University of Rochester Press, 2013); Joel Braslow, *Mental Ills and Bodily Cures: Psychiatric Treatment in the First Half of the Twentieth Century* (Berkeley: University of California Press, 1997).

7. See "The Brain in a Vat," ed. Cathy Gere, special issue, *History and Philosophy of Science Part C: Studies in History and Philosophy of Biological and Biomedical Sciences* 35, no. 2 (2004), and especially within this issue Charlie Gere, "Brains-in-Vats, Giant Brains and World Brains: The Brain as a Metaphor in Digital Culture," 351–66. See also Fernando Vidal, "Ectobrains in the Movies," in *The Fragment: An Incomplete History*, ed. William Tornzo (Los Angeles: Getty Research Institute, 2009): 193–211.

8. There was, of course, no single mind-body problem, but rather a set of questions about the relationship between consciousness and matter that was articulated differently at different times. Philosophical accounts of the relationship between mind and matter include: David M. Armstrong, *A Materialist Theory of the Mind* (London: Routledge and Kegan Paul, 1968); Daniel Dennett, *Consciousness Explained* (Boston: Little, Brown, 1991); Jerry Fodor, *The Mind Doesn't Work That Way: The Scope and Limits of Computational Psychology* (Cambridge, MA: MIT Press, 2000); Saul Kripke, *Naming and Necessity* (Oxford: Blackwell, 1980); Roger Penrose, *The Emperor's New Mind: Concerning Computers, Minds, and the Laws of Physics* (Oxford: Oxford University Press, 1990); Karl Popper and John Eccles, *The Self and Its Brain* (New York: Springer, 1977); Gilbert Ryle, *The Concept of Mind* (London: Hutchinson, 1949). On the longer history of the philosophical mind-body problem, see Tim Crane and Sarah Patterson, *History of the Mind-Body Problem* (London: Routledge, 2000). See also philosophical works drawing on embodied cognition to circumvent problems of dualism, e.g., Maurice Merleau-Ponty, *Phenomenology of Perception*, trans. C. Smith (1945; London: Routledge and Kegan Paul, 1962); Francisco Varela, Evan Thompson, and Eleanor Rosch, *The Embodied Mind: Cognitive Science and Human Experience* (Cambridge, MA: MIT Press, 1991); James J. Gibson, *The Ecological Approach to Visual Perception* (Boston: Houghton Mifflin, 1979); Alva Noë, *Action in Perception* (Cambridge, MA: MIT Press, 2004). For an articulation of the historicity of the mind-body problem, see Larry Sommer McGrath's *Making Spirit Matter: Neurology, Psychology, and Selfhood in Modern France* (Chicago: University of Chicago Press, 2020); and Joshua Bauchner, "The Lives of the Mind: Scientific Concept and Everyday Experience from Psychophysics to Psychoanalysis" (PhD diss., Princeton University, 2021). For histories of transparency, both metaphorical and real, see Stefanos

Geroulanos, *Transparency in Postwar France: A Critical History of the Present* (Stanford, CA: Stanford University Press, 2017); Daniel Jütte, "Window Gazes and World Views: A Chapter in the Cultural History of Vision," *Critical Inquiry* 42 (2016): 611–46.

9. Cecilia Heyes and Caroline Catmur, "What Happened to Mirror Neurons?," *Perspectives on Psychological Science* 17, no. 1 (2022): 153–68.

10. René Zazzo, "La genèse de la conscience de soi (La reconnaissance de soi dans l'image du miroir)," in *Psychologie de la connaissance de soi, Symposium de l'association de psychologie scientifique de langue française (Paris, 1973)* (Paris: Presses universitaires de France, 1973), 145–213, on 146.

ARCHIVAL DOCUMENTS

Papers of Hilde Bruch, Manuscript Collection No. 7 of the John P. McGovern Historical
  Collections and Research Center, Houston Academy of Medicine, Texas Medical Center
  Library, 1133 John Freeman Boulevard, Houston, TX 77030.
Archives of the Burden Neurological Institute, Science Museum Library and Archives, Science
  Museum at Wroughton, Hackpen Lane, Wroughton, Swindon SN4 9NS, UK.
Darwin Correspondence Project, Cambridge University Library, West Road, Cambridge CB3
  9DR, UK, www.darwinproject.ac.uk.
Darwin Online: darwin-online.org.uk.
Dr. G. Stanley Hall Collection, Archives and Special Collections, Robert H. Goddard Library,
  Clark University, Worcester, MA 01610.
The National Archives at Kew, Richmond, Surrey TW9 4DU, UK.
Papers of the Ratio Club, GC-179 B, Wellcome Trust Archives and Manuscripts, 183 Euston
  Road, London NW1 2BE, UK.
Papers of Edmund Snow Carpenter, Rock Foundation, 222 Central Park South, New York, NY
  10019. The papers are now at the National Anthropology Archives at the Smithsonian
  Institution.
Washington State University Archives, Washington State University, Pullman, WA 99164–5610.

INDEX

A NOTE ON THE TYPE

This book has been composed in Arno, an Old-style serif typeface in the classic Venetian tradition, designed by Robert Slimbach at Adobe.